THE SECOND COMING

The mighty ones of eternity

Roué Hupsel

AuthorHouse™ LLC
1663 Liberty Drive
Bloomington, IN 47403
www.authorhouse.com
Phone: 1-800-839-8640

Published by AuthorHouse   04/25/2014

ISBN: 978-1-4918-7162-1 (sc)
ISBN: 978-1-4918-7164-5 (hc)
ISBN: 978-1-4918-7163-8 (e)

Library of Congress Control Number: 2014904654

ESV
Unless otherwise indicated, all scripture quotations are from The Holy Bible, English Standard Version® (ESV®). Copyright ©2001 by Crossway Bibles, a division of Good News Publishers. Used by permission. All rights reserved.

Photo Roué Hupsel: Henna Brunings—Art Design, Paramaribo

PROLOGUE

Siloam Springs, March 4, 2016

He shuddered all over again when, for the umpteenth time, he studied the mysterious text in *Cuneiform script* on the tattered clay tablet.

"The soul eaters from the dark universe will come down on the earthly humans like hungry vultures and take possession of their bodies—woe to those who cannot escape . . ."

He shook his grey head in despair and stared again, captivated by the two images, depicting two planets. He took an antique magnifying glass from a tray to have a better look. On one of the globes something very small was written in *Cuneiform*. He was startled, shuddered again and became very excited by what he read. The most fantastic tales about this planet had been circulating among his colleagues and others. He sat down, staring silently in deep thought.

That planet was, therefore, real and not a fabrication!
And what about that mysterious text?
Was it a foreboding?
A warning?

A chill ran down his spine. He stood up again. His skeptical intelligence struggled to simply accept what was written there. Deep wrinkles appeared on his forehead. He clenched a fist on the armrest of his chair and relaxed it again. After hours of pondering and philosophizing, he decided to go to bed and have his two assistants involved the next day—maybe those two could shed light on this.

He examined the tablet with Don and Carl. They studied and researched it meticulously, presented the necessary caveats, accompanied by heated discussions, while constantly interrupting one another. After a full day of thorough research, something strange occurred during his sleep that night. The dream took place in the room where the tablet lay on a table. He dreamt that he was staring at it and in some strange way, was slowly being pulled towards it. The planets depicted on the tablet began spinning in dazzling bright colors. The room faded and dissolved in those colors.

Suddenly it seemed as if everything stood still, for a brief second . . . frozen and rigid. It was a kind of deliberate silence, threatening and frightening. Then very slowly . . . the distorted face of a very old man, dressed in a long black cloak, read aloud the text engraved on the tablet, in a language full of strange sounds. The old man turned his head towards him and looked at him mockingly with his hollow eyes. The old man's evil face, carbon-black eyes above an aquiline nose, a small mouth and a pointy chin like the tip of a lady's high heel—suddenly began spinning like a vortex . . . faster and faster, in increasingly small circles, accompanied by eerie, demented and hysterical laughter . . . . It resolved in colors becoming continuously darker, until there was total darkness. From this pitch-black darkness, something began bubbling up, then something menacing rose from the dark, encapsulating him like an asphyxiating cocoon. This lugubrious nightmare exposed him to truths not susceptible to the ordinary human mind. He struggled to get out of this nightmare and woke up completely terrified. Overcome with an indefinable fear, he looked around the bedroom, trembling and fearing a vague lurking mystery he could not grasp with his common sense.

The clock on his night stand showed 04:45.

Exhausted, sweating profusely, tormented by an intense thirst and with an enormous headache, he got up and stumbled to the kitchen to have something to drink.

These exhausting nightmares, in many terrifying variations, repeated in the next few nights and brought him close to the brink of madness. He sent for a doctor, who could only diagnose that he was suffering symptoms of extreme exhaustion, probably due to lack of sleep. He was hospitalized urgently. At the hospital, he fortunately did not have any nightmares. However, his condition deteriorated so rapidly that he slipped into a coma.

A few days passed and one afternoon he awoke with a frightened look in his eyes. He began speaking softly. His son leaned over, to understand the words that came from his father's trembling mouth. "*They are on their way . . . They are coming . . . This . . . This will be the worst disaster that will befall humanity . . . Their ultra-modern world . . . They . . .*"

The exhausted professor closed his eyes. Deeply concerned, the doctor approached the bed. With questioning eyes, he anxiously looked at the doctor, who shook his head. According to the doctor, the professor's feverish mind was apparently occupied with strange and incomprehensible things.

Suddenly, the patient opened his eyes. He began talking again, slowly and hesitantly. "*They were already on earth, in ancient times and brought us civilization. Now thousands of years later, they are physically weakened and have hugely decreased in numbers . . . They . . . They . . .*"

Once again a frightened expression appeared in the watery eyes of the old man. His parched lips started muttering. Tears welled up in his grey eyes. Then the old professor told an incoherent tale, the son did not understand, but which evoked his enormous interest nonetheless.

The scholar closed his weary eyes.

The worried doctor leaned over his patient.

The patient began speaking again. "*I have deep pity for humanity . . . They are coming . . . They are coming . . . We are beyond redemption . . . They . . .*" He made a desperate, heartbreaking effort to say something again. With his eyes he begged to be understood. He opened his mouth, but only a slight groan escaped his dry lips. Dave tried to comfort him, talk to him. The professor tried his best to answer all the questions raised. His lips moved, but there was no sound. Dave was not sure whether his father understood.

The scholar blinked.

He coughed twice and rattled briefly.

It was suddenly dead silent.

The doctor felt the patient's emaciated pulse and pressed the stethoscope against his chest. "He's gone," the physician whispered and from the corner of his eye, he saw Dave, standing there, completely stunned clenching his fists in helpless rage and sorrow. His thoughts going frantically to and fro!

These events started six years ago, when an American civil engineer happened upon a small, slightly damaged clay tablet while excavating an old pipeline in Iraq. The tablet, like all the other thousands found previously, consisted of flattened cushion-shaped clay, upon which the scribes had drawn horizontal and vertical lines to form squares or rectangles, within which pictograms were etched. This was done with obliquely cut reed stalks. The tablets were then baked hard in the hot sun. He managed to smuggle the ancient object into the U.S. Two weeks ago he showed the clay tablet to the professor, whom he had met at the University of Northern Arizona.

Dave too had seen the clay tablet with the strange *Cuneiform* inscriptions, when he surprised his dad with a quick visit during a short vacation. Robert Long, PhD, Professor of ancient extinct languages at the Northern Arizona University was an expert in the interpretation of the complicated *Cuneiform script* of the *Sumerians*. What Dave noticed when he looked at the clay tablet, was a drawing depicting a number of stars and two circles connected to each other by a line. His father interpreted the mysterious caption. Dave pointed to the pictured stars and circles.

"And what are those stars and two circles?" A deep frown appeared on the scholar's forehead. He scratched his chin and looked at his son with a faint smile.

"That is clearly the image of two planets. I need to study this clay tablet more thoroughly. It seems like a very interesting find."

And so he did, with disastrous consequences for him.

Dave Long spent almost two hours talking to the two assistants, who helped the professor with the examination of the clay tablet. Both told a very disturbing story. During the time the professor was in a coma, they too had experienced some bizarre dreams. The intensity of the dreams had been powerful. Carl Kopinsky, one of the assistants, admitted he was terrified by those emaciated beings that approached him. One of them had made desperate attempts to penetrate his mind. At least, that's what it felt like. He had awoken screaming and trembling all over his body.

Dave could find no explanation for the fact that both men, after examining the clay tablet, had experienced terrifying nightmares. He thought of his father, who fell into a coma and before his death spoke of sinister events, which sounded so incomprehensible . . . "*They are coming and we are beyond redemption . . .*" These words still sounded in his head. He decided to have another conversation with both men after the funeral.

Quietly he looked at his own reflection in the glass of the window pane in front of him.

Suddenly he was overcome with a deep rising fear. A fear, he was sure, would never leave him . . .

ONE

(Episode 1)

And it came to pass, when man began to multiply on the face of the land and daughters were born to them, the sons of God saw that the daughters of man were attractive. And they took as their wives any they chose . . . The Nephilim were on the earth in those days and also afterward, when the sons of God came in to the daughters of man and they bore children to them. These were the mighty men who were of old, the men of renown.
Genesis (6:1-4)

The gleaming white *starship* glided from the dark space of the universe, circled twice around the planet, and landed with a thin whistling sound on the ground, covered with wild grasses.

The journey from hyperspace had barely lasted four hours. The starting point of the *starship* was a planet, which after a very long period of time, had emerged from the shadows of the big Neptune and had moved progressively closer to Earth. A quarter *annum* ago they had left their world behind, with destination this blue planet.

The blue planet was not only one of the youngest of the ten planets circling the sun, but also the only one with flora and fauna strongly resembling their birth planet's.

Previous studies had also clearly indicated that a creature roamed there, that only several million *annums* ago had made the arduous ascent to developing a slight self-consciousness.

The vegetation was wild and grew disorderly high.

Beneath the flat land, hills and mountains, nature formed numerous rich minerals, among which a precious yellow metal, they were in dire need of on their own planet.

Through a complicated process of conversion to an electrospray of ions, it was the only resource to help keep their atmosphere intact and prevent the terrible rise in temperature. Without that yellow metal, life on *Nibiru* would be irretrievably lost. An awful lot was depending on this mission, which could not fail . . .

Anxious and quiet they hid behind the wild growing bushes. Three men, four women and one child trying to process with great difficulty what they were staring at. It was a late afternoon; the sun colored the landscape brilliant red. Slowly, majestic, sparkling white and painful for their astonished eyes, reflected by the rays of the setting sun, the enormous star broke through the clouds and descended slowly, accompanied by carols and other heavenly sounds.

Moments later, the impressive mighty star touched down on the scorched grass.

Suddenly it was quiet.

Then an entrance to the mighty thing silently opened.

Three divine creatures appeared, tall and proud. The first one, a wonderful goddess, looked around. Instantly her eyes focused on the spot where they were. Respectful and frightened the men, women and the child fell to their knees, overcome with tremendous fear and an indescribable, unprecedented emerging emotion.

This was a moment they would never forget.

The gods had descended from the heavens . . .

Mesopotamia, the land between the rivers

Thirty thousand years B.C.

> *And they had not set up a King for the beclouded people. No headband and crown had been fastened; no scepter been studded with lapis lazuli . . . scepter, crown, tiara, headband and staff were still placed before Anu in heaven . . . There was no counseling of its people; then the Kingship descended from heaven.*
> (Clay tablet from the library of King Ashurbanipal)

The wise *Nin-khursag, Lady of Life* was in the temple of glorious *Nippur*, thoughtfully staring at the wall on which *"the Table of Destiny"* was displayed. Not only the story of their arrival on Earth, the events that subsequently took place, the course of the planets, but also other words of wisdom were described in flamboyant *hieroglyphs*.

The humans with the frizzy hair were not privy to the secrets of these hieroglyphs—for they were the result of the great experiment. They would be taught *Cuneiform* at a later stage.

But the experiment with these humans had borne bitter fruits. The dark frizzy haired humans of *Naphidim* had become a threat to the other tribes living around them, due to their weapons based autocracy and dominant culture. They were also bigger and stronger than the members of the numerous *Eljo* tribes and the other savage races, which they tried to destroy completely. *Nin-khursag* closed her eyes, opened them again staring with a dull gaze, as a sour smile appeared on her lips. It was to be expected—the women of *Naphidim* were extremely attractive and numerous *Anunnaki* had sired children with these beautiful women. She had watched this with an alarming sense, especially when she noticed that some cases threatened to get out of hand. Attempts by males of her species, to try and change the minds of the sons they had fathered with these women, were without success. It was all in vain. These findings not only made her sad, but she also became very unsure of how their mission on this planet would end.

> *Many angels of God accompanied with women, and begat sons that became unjust . . . on account of the confidence they had in their own strength; for the tradition is that these men did what resembled the acts of those whom the Grecians call giants.*
> (First century Jewish antiquities: Flavius Josephus)

Nin-khursag was also deeply concerned, for in this very temple *The Great Assembly of the Anunnaki,* in the Court of the Highest, had adopted the decision to end what they had created many *annums* after their arrival on this planet. *The Great Assembly headed* by *Enlil* had unanimously taken the decision to destroy. The solemn *"haem" "so be it"* sounded in her head, as she slowly stood up and walked forward. Her green eyes with the long black lashes looked straight ahead.

A primary function of the *Assembly* was the appointment of kings, but it was also a court of justice. But that had been reduced to something that was totally not important at this moment.

A faint smile appeared on her face when she thought of the name *Anunnaki "Heaven came to Earth,"* that's what the *Naphidim* called them out of reverence, respect and fear.

She got a sad look on her face, when the *High Assembly's* recent decision entered her mind again. The beautiful fertile land of *Sumer* would soon be covered with icy cold water.

> *Man of Shurrupak, son of Ubar-Tutu,*
> *Tear down thy house; build a ship,*
> *Abandon thy possessions, and seek thy life.*
> *Discard thy goods, and keep thee alive.*
> *Aboard the ship take the seed of living things.*
> (Library of King Ashurbanipal)

She slowly rose and strode into the courtyard. She paused with a pensive look in her eyes. When she turned around to follow a beautiful blue butterfly, flying from flower to flower, she caught a glimpse of the reflection of her face in the polished black stone door leading into the temple.

Nin-khursag, the *Lady of Life,* was a beautiful divine appearance. Her face with the serpentine features, small forehead, high cheekbones and almond-shaped green eyes, was framed with wavy black hair reaching to her waist. She had soft light brown skin matching her long dress contrasted with rose and gold. *Nin-khursag* was slim with round hips and she measured almost 8 feet tall.

Under heavy protest, she had voted in favor of the motion. She disagreed with the decision and regretted it. She also understood, however, that *Enlil* and the *High Assembly* were at their wits' end. With selective famines and plagues they had tried to curtail the rampant population growth, but that had not helped.

They had not succeeded to impart the key elements of wisdom and teach these humans to live in a well-regulated society. They did not know the institute of a man taking a woman and remaining faithful to her the rest of his life. Sexual intercourse with these *earthlings* was a matter of free will, with a freedom of partners. Disorderly they dwelt in their big cities, where they held the largest and most vulgar orgies, accompanied by an awful lot of noise, to the chagrin of those they called gods.

If there were another plan after the Flood, it would be of the utmost importance that the new *earthlings* be conditioned in such a way, that they would be in strict compliance with municipal law and accepting of well-organized social governance.

Deep in thought, she slowly strolled through the courtyard and watched her shadow getting shorter, as the sun reached its zenith. The fragrant aroma of blossoms was all around her. She savored it with every breath, sighing audibly and again a pensive look appeared in her eyes. She stood awhile, enjoying a few moments of tranquility around her. Then she turned around and slowly strode back to the entrance of the temple. She sat down on one of the sheepskin sofas and folded her hands.— There had to be a second plan, there was no doubt about that, but for everything to work smoothly this time, they had to properly consult on a flawless execution thereof. Her contemplation was interrupted, when she heard the sound of light footsteps behind her and a deep baritone voice: "We need to talk, Lady."

When she turned around, she saw *Enki the Wise,* the brother of *Enlil,* with a serious look on his face, walking towards her at a uniform pace.

"I have been expecting you one of these days", she said and her full lips formed a faint smile.

"Yes, my Lady, I have been worried since the meeting. The program did not go as we expected. This breed is totally depraved." "What did you expect?" she said. The tone in her voice clearly betrayed her disappointment. She chose her words carefully. "They multiplied beyond any control and the degeneration occurred just as rapidly. Mankind did not turn out the way we had hoped." She shook her head and for a quick second a dull look appeared in her eyes. He reached out and touched her arm.

"There are also individuals who are decent, civil, and who try their best to live righteously and behave the way we really wanted, but . . ."

He abruptly stopped talking.

"Go on, I'm listening," *Nin-khursag* said softly, playing with a lock of hair.

Suddenly, his voice was resolute.

"I have a plan!"

"And? Tell me Wise One. I'm curious."

Enki paused and looked deep into her eyes.

"I am taking measures to save one of their kings."

She gasped and looked at him questioningly.

"What's the name of the king?"

Enki pulled nervously at his beard. He took a deep breath and said softly: "*Shurrupak.*" "Hmm . . ." *Nin-khursag* nodded surprised.

"Why him?"

"I have given this a lot of thought. He is the best choice. He is a quiet and wise man, who for a long time has been dissatisfied with the conduct of the humans. He has always behaved in an exemplary manner. If only the others were like that too."

Enki fidgeted with his robe and shook his head to hide the slight discomfort that suddenly came over him. He shrugged and looked at her again. "They have a free will, which in the end led to their demise. That's the demise that is yet to come, and of which they are unaware. *Shurrupak*, the exception among the masses, has given me a glimmer of hope that our mission on this planet still has a chance for success."

"And how would you save him?" *Nin* asked. She looked at him expectantly.

"Come with me." His voice was tense. They exited the temple and entered the courtyard. The sun cast its fiery rays through a dust lane in the courtyard and illuminated the branches of a mohu tree that stood as a stable union between the lush greenery of the mahai plants. She sucked in the cool air with deep breaths. They sat down on one of the benches under the tree. He stared in the distance for a while, and then started talking. "I plan to have him escape by means of a specially built boat."

"Oh, does he know that?"

"No! Not yet. I have an appointment with him tomorrow afternoon."

"And what kind of boat will it be and what are your further plans when we have the water come down?" "Incidentally, *Enlil* has already given the orders. *Ninurta* and *Enugi,* entrusted with that task, will soon make the necessary preparations." Her face tightened and a shiver went through her shoulders. She recovered quickly and looked at him with a slight smile on her lips. He understood her mood perfectly well, but pretended not to have noticed anything.

Softly she said: "Continue please."

"As I was saying, it will be a special boat, where the seeds of mankind, plants and animals will be stored."

"Thus *Shurrupak* will not only be the guardian of human seed, but also of the seedlings and seeds of plants and animals?"

"That's the plan."

"I need your assistance to prepare these seeds and place them in the stabilizers."

"I will be pleased to comply with your request," *Nin* responded with a beaming smile.

"I did not expect a different answer from you *Lady of Life.*"

Lazily and in a better mood she leaned back on the bench.

"Tell me more about the boat."

He clasped his hands. A proud look appeared in his eyes.

Enkidu spoke thoughtfully and enthusiastically: "It will be a boat that is impervious to rough water and will be able to maneuver well above as well as under water. This sealed vessel made from the shiny white metal that we extract from the red rocks, will perpetuate my plan to preserve some of what we will soon have destroyed."

"But I wonder how my spouse and brother *Enlil* will react to these plans." She looked concerned.

"We will worry about that later," *Enki* sighed. He slowly rose. I am leaving now—there is so much to do, for you too."

"I will start as soon as possible with the creation of the seedlings."

She rose also and they did the laying on of hands.

Then, deep in thought, they exited the courtyard.

TWO

Four weeks later . . .

The weather had been alternating for several days, some rain and a lot of sun. That was normal for this time of year. But that could change very soon. This morning some rain fell. During the rest of the day, it was beautiful sunny weather; a blue sky, here and there fleecy clouds slowly moving by, and slowly changing shape. But that would not last much longer. *Shurrupak* peered at the sky. Would he make it? He had to hurry. The days flew by quickly. *Enki* came to see him a few weeks ago, to deliver the package with the drawings of the boat.

He was surprised, when a few days earlier a messenger had handed him the invitation to visit *Enki* at his palace. No, he did not expect it. Curious about the reason for the invitation, he headed out immediately. *Enki* gave him a carefully worded explanation of what lay ahead. *The High Assembly of the Anunnaki* had decided to end the sinful, noisy humans with the power of water. With bowed head, he had respectfully and earnestly let the words wash over him.

"I chose you because of your lifestyle and your attitude to surviving all this, so that not everything we created will be lost."

They spent many hours discussing the entire plan in minute detail.

The construction of the submersible vessel, which was completely sealed, had gone according to plan. The white shiny metal dolls, programmed to execute very specific tasks, accomplished their jobs on time. The clinical container vessel looked beautiful, yet functional.

Nervously, he awaited the arrival of *Nin-khursag*, who handed him the generated seeds and ova contained in the chilled cubes.

Seven weeks later . . .

He took one last look at the land that would be completely submerged. The metal dolls had been taken away by *Enki* after their work was done. The seeds and ova were securely stored in cool areas of the boat. He had also brought on board an ox, four sheep, two rabbits and a few canaries. He had done so, after *Enkidu* had smilingly granted permission to his hesitant request.

Many had mockingly watched the construction of the boat.—*Who would build such a thing, far away from a river and the sea?* He had not reacted to the scornful laughter. Calmly he had continued with the construction, assisted by the silently moving metal dolls that knew exactly what to do.

From the chronicles of Shurrupak

Then . . .
In the middle of a beautiful sunny day it began!
The sunlight fades to a pale glimmer that suddenly turns into a creepy kind of nasty dark grey.
Within a few minutes the weather completely changes. A cool breeze turns into a violent squall, blowing harder and harder accompanied by a bone chilling, shrill howling sound.
There is thunder and lightning. Fiery tongues color the sky yellow, followed by loud blasts of impact. The lightning continues and it starts to rain. Never before has such a heavy downpour fallen from the dark grey sky.
Suddenly, it turns dark . . . the sun has disappeared!
The earth begins to vibrate . . . Vibrations that become stronger and turn into earthquakes.

The earthquakes cause the earth to shake to its foundation, over and over again.
In some places the shockwaves rupture the earth, opening up fathomless abysses, devouring the earthlings like a hungry monster.
Desperate, loudly screaming men, women and children run back and forth helplessly.
Some fall down on their knees crying, begging, and praying to the gods.
But it is too late . . .

After two weeks . . .
Excited, he stared once more at one of the monitors installed on the boat. Outside it was dark. Utter darkness prevailed in the daytime too, sunlight had disappeared completely. He was shocked by what he saw, when a fierce flash of lightning brightly illuminated the surrounding area for a brief second. His eyes narrowed. The wrathful water swirled around the boat, carrying on its fierce pounding waves the faint remains of humans, animals, uprooted trees, plants and other broken unidentified objects.

He was saddened that this was the end of a world that, notwithstanding the bad behavior of the humans, had looked so beautiful. He would surely miss the rolling hills, in some spots covered with tall shrubs, the green forests in the west, the grassy meadows and the snow peaked mountains to the north of where he had lived since birth.

All this was now disappearing rapidly.

The hungry, feet high crashing waves, accompanied by deafening lightning, turned the land into a fierce raging sea.

(From the annals of the starship)

After a few weeks it was all over.

Enlil's granddaughter *Inanna,* who was always in front of the monitors of the big starship, swiftly moved her fingers over the panels.

Outside the sun was at its highest point.

She suddenly let out a surprised cry. "Truly, the days of yore have turned to clay!"

The ground far below the starship was flat like a brown slippery roof, extending infinitely.

Via one of the monitors on the boat, *Shurrupak* was able to conclude that the weather was changing, after all those weeks of crashing precipitation. Since a few days, there had only been a misty drizzle and the sun hesitantly began to shine on the land again.

Then suddenly a beautiful rainbow appeared.

He smiled . . . it was over!

He decided to set one of the canaries free. If the bird did not return, it would be confirmation that somewhere it would have found dry land. Confirmation of that fact came when the bird did not return, after waiting in vain for several days. He continued sailing around a few more days.

Outside an island or two appeared with the beginning of hesitant growth of green grass. He chose one to dock the boat. It was a beautiful island, with lots of starting greenery between the clay.

The weather looked splendid again today. The past few sun-drenched days were followed by evenings with countless twinkling stars and a moon which, as far as he could remember, had never looked so beautiful in the sky. He was saddened, though, when he stepped off the boat and cautiously set foot on the soft clay.

The old world was gone. The silence around him was oppressive the first few hours, but he gradually grew accustomed to it and even found it enjoyable after a while. An optimistic smile appeared in his eyes, when he looked around again. Soon not only this island, but also the mainland would be populated and cultivated by the new *earthlings*.

His smile widened, but suddenly he looked serious again. He had to begin the big task very soon.

The next few hours he worked steadily on the preparation thereof. After a day he was ready. He touched the panels of the instruments and with a high buzzing sound everything came into operation. The seeds and ova were led through the stabilizers and placed in cubes, specially reserved for the generating process. Cloning and other ongoing business would follow later.

The new *earthlings* would be prepared for their tasks, when this mighty work was completed. In their genes, the necessary *Anunnaki* cells had been placed in order for the conditioning to proceed favorably and positively.

These new humans also had to be given a clear explanation of what "had occurred" before they were created.

THREE

Nin-kursagh, who is of unique greatness,
Makes the womb contract.
Nin-khursag, who is a great mother,
Begins the birthing process.
(Clay tablet from the library of King Ashurbanipal)

Angrily *Nin-kursagh* flung her heavy silver crystal-inlaid goblet, filled with precious wine, to the ground. The shards flew everywhere and a red wine stain appeared on the floor. Once again she was embroiled in a fierce debate with her brother and spouse *Enlil*, which usually ended in a huge fight. Furious, she walked back to her luxurious cabin, where she grudgingly sat down on her sofa. She closed her almond-shaped eyes with the long black eyelashes and took a deep breath. Slowly her anger subsided, after she slightly touched the pleasant-mood panel with one of her fingers.

The fact that they would need her was beyond doubt. Her work was sacred to her. No one on her own planet could compare to her. She was born for this. Musing she looked ahead.

She also thought of *Enki,* who supported her in her work and who always stood by her with wise counsel, whenever she asked for it—she could always turn to him for advice.

As she lay there pondering, she got in a better and more pleasant mood. A satisfied smile appeared on her luscious lips, when she realized the task ahead and its full extent . . .

Nin-khursag was a highly regarded anatomical specialist. She was also responsible for the stored semen of the wise *Enki*, used for cross-fertilization of other life forms on this planet. Her lab would go down in history with the *earthlings* as the house of *Sjimti*—breath, wind and life. Her experiments were carried out with increasing accuracy and reached unprecedented perfection.

She had a big dream and was all set to create her masterpiece, the ultimate human being—*the Homo sapiens sapiens!*

Enkidu entered his cabin and pulled his purple robe tight. He sat down on a sofa and stared pensively into space. He was worried about *Nin-khursag*. The persistent arguments and quarrels between her and *Enlil* endangered the defined objectives they intended. The two of them got along well at times, but not for long, especially when they indulged in much wine. They hurled the bitterest accusations at each other. Trivial matters were vastly exaggerated.

From the very beginning, *Enlil* had objected to educating the *earthlings*. He considered them mostly laborers, who had to work the mines and serve them. He was the one who always favored to cracking down on the noise they made in the cities and their dissolute way of life. He also was the one who, assisted by his two advisors *Ninurta* and *Ennugi*, invented the plagues and famines and also conducted those, in order to mercilessly bring the ever increasing mass of *earthlings* to acceptable proportions.

In contemplating these matters, a disappointed look appeared in the eyes of *Enki the Wise*. The first experiment was a complete fiasco, but he was hopeful. Tomorrow he had another important meeting with the *Lady of Life*. A lot was pending on the outcome of that conversation. He trusted and relied on *Nin* and eagerly looked forward to tomorrow.

Enkidu was over eight feet tall. Intelligence radiated from his light green eyes under the thin black eyebrows. His powerful mouth and beard gave his face an alternating mild and stern expression. His black hair hung to his shoulders in *Anunnaki* style. His broad chest and narrow hips attested to masculine power, which he could not hide under the long purple robe. He descended from the *Nanas,* a very long time ago the dominant race populating the southern part on the home planet. *Enkidu* was proud of his roots and of himself.

The ancestors of the inhabitants on his home planet had ensured a rapid artificial evolution of the race. *Annums* ago, after many experiments, they had also drastically extended by many hundreds of *annums* the lifespan of males and females. All illnesses belonged to the past as well. After the diseases had been eliminated, the species grew into healthy superior beings, elevating science to unprecedented heights. However, they could not attain immortality. That had proved impossible and irrelevant.

A major step forward was exploration of the universe. This resulted in many disappointments, because many planets they visited proved uninhabitable for them. Huge and surprising was the discovery of a new world resembling theirs. The psi contact the inhabitants of that planet had sent forth was not great, but that needed further investigation. The result was that, after many *annums,* they landed on this planet with much optimism.

His cabin on the *starship* was comfortably furnished with soft sleep sofas. The two large monitors were turned off. The artificial lighting responding to his mood was on medium. Slowly he got up and walked over to his sofa bed. He lay down comfortably on the soft cover. The power turned on automatically and caused him to quickly doze off in a refreshing sleep.

FOUR

Nin-khursag and *Enki* sat across from each other and discussed what had to happen after the flood. *Enki* looked serious. "We have no time to lose. The once fertile land has turned into a large plain of clay and the entire environment has been destroyed. We have to start all over again." These words were followed by a deep sigh. He looked at her hopefully.

Nin straightened her back and a radiant optimistic smile appeared on her face.

"Our first priority is to make the soil fertile again, and for that we need strong healthy *earthlings*."

"Yes, *Lady of Life*, and you will get yet another great roll assigned therein."

"Yesterday I had the necessary sheep and bovine ova removed from the cubes in the growth chamber and placed in the creation room. I also perfected the breeding area for rapid growth. In a few days the process will be completed." *Enki* nodded approvingly. "*Nin-khursag, annums* ago you were the best choice *the Assembly of Twelve* on our home planet made."

"I have also given some thought to the next phase. In order to accelerate the execution of this process, I will need female *earthlings* who have survived the deluge."

"I have good news for you, a few survivors were found on an island. Obviously, that island was a high mountain before the disaster."

"Yes and that island has become again what it once was—a mountain."

"How many survivors are we talking about?"

"Three women and a child—a girl."

Nin-khursag nodded, satisfied. "Excellent, we shall proceed with the nigh culmination of the experiment."

"They will be brought to you tomorrow."

"Fine, the sooner the better."

She looked at him.

"In this second important phase I will add, as planned, more of our own genes to the cross-fertilization. This has to become a totally different kind—a species that can be taught and socially regulated."

"The agreement you and I previously made, hopefully, will completely succeed now!"

"This time it will definitely be alright. I have gained enough experience. And my latest experiments are very promising."

They were silent awhile, facing each other.

Then they did the laying on of hands and *Nin-khursag* slowly walked off with a smile on her face.

She got in an increasingly positive mood after *Enki's* visit. After tomorrow she could continue the great work—create life! That last thought made her chuckle. She did not create life, but she gave a big upward push to the natural process of creating a new human. She also thought of this human race that she was in the process of cultivating. The *earthlings* had an exaggerated respect and reverence for those who had come from afar. They considered them gods from heaven who had descended on this planet. In their shortsighted metaphysical thinking, this resulted in great errors, misunderstanding and confusion.

She untied her hair, shook her head and luxurious locks fell to her shoulders. During her work in the lab, her hair was always hidden under a cap she turned on with thought impulses. *Nin-khursag* allowed nobody in her lab—it was off limits to everyone. She received *Enki the Wise* outside. He understood completely and also wanted her to carry out her work undisturbed. She had access to the best equipment to optimally do her job. After being selected by the *Assembly of Twelve* on her birth planet, she underwent rigorous training, resulting in her official appointment as *"Lady of Life."* She was appointed to that position shortly before the *starship's* departure, with destination the young planet, part of the solar system. The wise ones had discovered an awakening consciousness of the lost and destructive creatures wandering there.

It all began when on her birth planet, emissions of psi powers of a lower order were observed at a planet on the edge of a spiral nebula. In the beginning, this phenomenon was perceived as unimportant. It appeared to them that it had barely transcended the animal level. Much later, they discovered that flashes of applied psi were received. They became curious and started scanning the entire planet for psi. They were surprised to witness the birth of an emerging civilization. To their disappointment, a standard message broadcast went unanswered. There was no response either to other messages sent. An ecological team with physical resources dispatched to investigate, discovered that without outside help, that planet's early life form could not meet the requirement for achieving rapid development of intelligent life.

This third planet in the planetary system of the tenth order aroused their interest ever more. Further expeditions were dispatched with small airships to study the creatures there more thoroughly. Study of the soil condition, flora and fauna were also part of these explorations.

In the universe, which is populated with an incredible variety of conscious life, there are clear common characteristics of planetary conditions making that life possible. A protective layer surrounding the planet has a filtering scatter effect, which is very important for the basic shapes that are the foundation for the development of life forms. It was observed that this protective layer consists of gasses and liquids to reflect the rays of the mother sun and other suns nearby, filtering and converting them until they are able to sustain life.

Contact with the beings existing there was very difficult. They either fled or threw themselves at their feet, as if they were the creator. It was an extremely wild and primitive creature that wandered there and lived from hunting and fishing. They lived in huts made from leaves of the luxuriantly growing plants. A hesitant beginning of speech and the use of objects made of clay, stone and wood was also detected.

The similarities with their world were decisive in the choice of that planet.

A livable climate for them and, most of all, in the soil that precious yellow metal, they were in dire need of, to further developing numerous advanced techniques in order to stem and eliminate the warming of their birth planet.

The decision was made to construct a new sophisticated *starship* and dispatch it to the small blue planet. They had gained extensive experience with the instruments used to detect and mine the yellow metal. There was a dramatic shortage of that precious commodity. The soil on their planet was completely depleted. Hence, the discovery of this blue planet was a gift of immeasurable value.

As the plans progressed, it was suggested to involve the most intelligent species on that planet in the labor process. But first, something had to be done about the thinking of those beings—transforming them into active workers. That is when she, *Nin-khursag*, entered the picture. She would be assigned the task of doing something positive for the physical and psychological nature of the inhabitants of this world.

The preliminary program was a success. But as the humans grew more intelligent, so did the degree of their degeneration. They multiplied at a staggering rate, became unruly, rebellious and boisterous.

Nin-khursag was opposed to destroying them, but the *Great Assembly of the Anunnaki* decided differently and she had to acquiesce.

After the deluge, there was a second opportunity to experiment with more know-how. With better material and new techniques, she planned to create better human beings than before.

At least, that is what she was hoping for.

She lay down on her sleep sofa and with her fingertips she swiftly touched a panel to recharge her body with new energy. Tomorrow promised to be an exciting and eventful day.

FIVE

Mother Nintur (Nin-khursag), the lady of form giving,
Working in a dark place, the womb.
To give birth to Kings, to tie on the rightful tiara;
To give birth to lords;
To place the crown on their heads.
It is in her hands.
(Clay tablet from the library of King Ashurbanipal)

Enki had to talk to *Nin* often after the demise of the *earthlings*. She was stubborn and proud of the work she did. Now that she was in complete accord with him, and after having had numerous conversations with her, he was able to look forward to a bright future.

Quiet and reserved, he accompanied the survivors to *the Lady of Life*. The women and the child fell on their knees, respectful and fearful.

"They look great indeed, excellent specimen," she whispered softly with an excited voice to *Enki*, who watched the *earthlings* with a smile.

They have been examined medically and have been found healthy for "*treatment*," he said encouragingly.

The women and the child were escorted by two metal dolls, to a special room with sofas.

"I will start immediately with the *treatment*, I have prepared everything already."

"Then I will leave you alone," *Enki* responded, with a slight smile.

"Good luck! I will follow your progress on my monitor."

"Please do, I will send you an update every other day."

After this promise, they did the laying on of hands. He walked with slow steps and in a good mood to his cabin.

> *The plow and the yoke he directed . . .*
> *To the pure crops he roared.*
> *In the steadfast fields he made the grain grow . . .*
> *Enkidu, him of the canals and ditches,*
> *Enki placed in their charge.*
> *The grains he heaped up for the granary . . .*
> (Clay tablet from the library of King Ashurbanipal)

Enki the Wise had big plans. But he could only realize those plans with the help of *Nin-khursag*. From the *earthlings* they had to develop a strong work force, to work in the mines and cultivate the land. New towns had to be built, but also a new social order had to be established, with *humanans* as the governors—in the first plan, these posts were occupied by his kind only. Of the utmost importance to reaching this goal, were the strict laws and order he would impose upon them. The innate savagery of this breed had to be eliminated through cross-fertilization and cloning until a new race should arise. Creating a new human that would meet their expectations, was a long process. But it had to happen—after all, the prospects were very promising, indeed.

Nin-khursag, who shortly after the flood, feverishly began working on the big assignment, was able to, after a few months, look back contentedly. She successfully created fourteen new humans: seven boys and seven girls. For this impressive process, she used the wombs of the women who had survived the flood. These wombs completed *Nin's* work, by developing the shape of the humans she created. The human ova fertilized by *Anunnaki* semen were transferred, in the form of cultivated embryos, into the wombs of these surrogate mothers. As a result, they were born quite normally as babies.

(From the Babylonian annals of Berosus)

A great number of laborers were created, to work in the fields and mines. They also began building new cities. In establishing the new social order, the carefully chosen leaders, called *humanans,* played a very important role. They were appointed governors and were in charge of the provinces into which the land was divided.

These humans, who called themselves *black heads* and who were the product of *Nin-khursag's* work, were also taught in their conditioning process that the main reason for their existence was to respect the *Anunnaki,* by providing them with food, drink and shelter. In return, they were taught and trained in social skills and academic affairs. Irrigation techniques and water management, division of the circle in 360 degrees, introduction of lingual systems, division of hours in minutes and seconds, denomination of the zodiac signs and the astrological system were successfully imparted unto them. Important to comprehending everything well, was the beautiful language they taught these *earthlings*, which the other tribes called "*Sumerian.*"

These new humans in *Sumer* abhorred evil, deceit, lawlessness and injustice. Their high priorities were goodness, truth, order and freedom. By the surrounding tribes they were, therefore, called *Kir-en-gir,* the land of the civilized rulers.

This was the period that would be known in the chronicles as the "Golden Age."

SIX

Oil he commanded for him, and he was anointed.
A garment he commanded for him, and he was clothed . . .
His command was like the command of Anu.
With wide understanding, he had perfected
him to expound the decrees of the land.
He had given him wisdom, but he had not given him eternal life.
In that time, in those years of the wise son of Eridu, Enki had
created him as a leader among mankind.
Of the wise one, no one treated his command lightly.
(Clay tablet from the library of King Ashurbanipal)

Enkidu and *Nin-khursag* were extremely satisfied with the results so far of these new *earthlings*.

"It is now time we advanced to the other plan."

Enki was in total agreement with her. "These humans need strong leaders, to guide them and to keep them on the right track. And how far have you progressed with those plans *Lady of Life*?" *Nin-khursag* smiled mysteriously.

"Very far. This time I will make my own body available . . . This new life will be created in my womb!"

He looked at her with big astonished eyes.

"Oh lady, there aren't any risks associated with this, are there?"

"Not at all, I studied everything thoroughly and have thought about it for a long time," she assured him. "It is my intention to place a cultivated embryo in my own womb, rather than in the womb of a mortal mother. I'm doing this in order to feed the embryo with my own cells and blood. I have developed an embryo that has been cultivated in the ovum of a mortal woman. She has been clinically fertilized with your seed." *Enki* looked concerned. *Nin*, who saw that, laughed away his worries. "It will be a successful experiment and the crowning glory of my work! It has to be a prototype for a race of superior earthly leaders."

After several months the time had come. *Nin-khursag* gave birth to a boy. He would be recorded as the Model of the *Earthlings*. *Enki* named him *Adapa*. This *Adama* earthling developed into a powerfully built, handsome young man, the pride of *Nin-khursag* and *Enkidu*.

Enkidu was so delighted with him that, after some time, he appointed him his personal delegate. In the city of *Eridu*, he was appointed head of the *Enkidu* temple in Sumerian *Eden*. He became the first ever priest of this planet.

> *With far-reaching understanding Enki perfected him*
> *to interpret the decrees of the land*
> *He had given him wisdom,*
> *but he did not give him eternal life.*
> (Clay tablet from the library of King Ashurbanipal)

Another important task *Nin-khursag* took on was the procreation of a life partner for *Adama*. This life partner was created in exactly the same manner as *Adama*. She was born from the womb of *the Lady of Life*. In esoteric language she would be called *Evan, Lady of all living*. The *Sumerians* called her *Nin-tî, Lady of Life*. *Nin-khursag* had big plans for her surrogate daughter as well.

(From the chronicles of Shurrupak)

The Ti (rib) symbolically denotes that she is equal to Adama. She is not meant to be ruled by him, for she neither grew from his head, nor did she grow from his feet to be trampled by him.

27

SEVEN

Plenty of the yellow metal was stored in the cargo hold of the large *starship*. With a melancholy feeling, *Nin-khursag* glanced a last time at the blue planet. Not only did she love life here, but she also loved the humans with the black hair who, in their sober reflections, did not regard or mistook them as gods. The *black heads* showed tremendous respect and reverence for those who had descended from *"the heavens"* and regarded them more as benefactors and founders, who had imparted unto them extensive knowledge of the things around them, as well as far beyond, such as the cyclical course of the distant suns and planets in the dark mysterious universe visible from Earth. Collectively and individually, they received a tribute they deserved—the *Anunnaki* were the founders and masters, who determined what was to happen in the land of *Sumer*.

(From the Babylonian annals of Berosus)

> *The black headed humans were called Sumerians by the other tribes, because of their domicile, the beautiful Sumer.*

The command to start the return journey came from *the Assembly of Twelve* on their home planet.

Nin-khursag was in a deep trance. In one hour the *starship* would begin the journey home. The preparations to blast off into hyperspace had been completed. With sadness in her heart, she looked at the monitors one last time. The memories of her sojourn on this planet would forever remain with her. She could not resist the urge to meditate shortly before takeoff. Her mind drifted slowly into a world of purple, red and light green . . . In a flash, she suddenly saw a future filled with beautiful color circuits which, to her horror, became increasingly dark.

The revelations came to her in a panorama of images.

Both races, the one on her home planet and that of this world, were connected by an invisible mystical bond, which would have dramatic consequences thousands of *annums* later.

What was revealed to her in those minutes was horrifying . . .

They, the inhabitants of *Nibiru,* after all those *annums,* would return to this planet and its humans . . . *to survive!*

After having been obscured for thirty thousand *annums,* the time would come again, according to the cosmic cycle that *Nibiru* would emerge from the dark shadows of Neptune and Pluto, this time with disastrous consequences for the *earthlings.*

For a brief second, she saw *Enlil's* hollow smiling face, contorted in an evil grin.

She also saw the final stage of her own species and the terrible outcome . . .

It was then she better understood the course of cosmic occult events.

All the physical worlds and the species inhabiting them were transient things.

They were flashes in an incomprehensible eternity.

Slowly she came out of this meditation.

Her eyes looked sad and the expression on her mouth conveyed a somber disappointing mood.

During the day it could be seen as a bright shiny silver dot and at night, it looked like a big twinkling star in the firmament.

Until one day, completely unexpected, the enormous *Heavenly Carriage* descended from the clouds, to depart after only a few days with *the Great Ones*. The severely disappointed black headed humans went in deep bitter mourning.

The mighty ones, who had created them and had bestowed upon them untold wisdom, had gone for good.

They had been left to their destiny.

(From the Babylonian annals of Berosus)

The encroaching barbarian tribes invaded the land of Sumer, with devastating consequences.

> *Ur is destroyed, bitter is its lament.*
> *The country's blood now fills its holes like hot bronze in a mold.*
> *Bodies dissolve like fat in the sun.*
> *The gods have abandoned us like migrating birds.*
> *Smoke lies on our cities like a shroud.*
> (Clay tablet from the library of King Ashurbanipal)

EIGHT

Episode 2

Siloam Springs, April 2016

He moved restlessly in his chair. Dave Long had retreated to the study of his father's house in order to think. He stood up, shook his head wearily and sat down again. His mobile phone vibrated in his pocket. He quickly pulled it out. A smile appeared on his lips, when he saw her face on the small screen and heard her little voice. It was his daughter, wishing him good night. "Good heavens," it suddenly occurred to him . . . "I forgot to call her". He usually telephoned her every evening, but due to the tragic events of late, keeping him constantly occupied, he forgot to do so this evening. After he finished talking to her, he stood up again. His thoughts shifted back to his father's passing. His father was a man, who seldom got sick, suddenly lapsed into a coma and died. His father's passing was a complete mystery to him. The last enigmatic words before he died caused a deep frown on Dave's forehead, every time he thought about it.

As the sole heir—Dave was an only child—he came in possession of all the papers, archive and computer files, which he had been studying for several days now. There was also a notepad, demanding his full attention. From his father's shoddy handwriting, Dave was able to decipher that he was referring to a planet yet to be discovered, which the clay tablet mentioned. The damaged clay tablet was proof that the ancient people had convincing knowledge of that planet. Those two spheres, representing two planets, kept him occupied. He knew the mysterious text by heart, at least the English translation thereof.

A great deal of the collection of his father's books dealt with the history of the mysterious humans, who lived many thousands of years ago, in the land of the two rivers. He delved into those books and was surprised to discover that so much literature had been published on the Sumerians. In one of those books, he also read that, thanks to a true library of well-preserved clay tablets left to the world by King Ashurbanipal of Assyria, so much knowledge of those ancient humans had become known. This king referred to relics of a great civilization, already 'prehistoric' before his time.

For posterity, he recorded in clear terms: *"The god of the scribes has bestowed upon me the gift of the knowledge of his art. I am privy to the knowledge of writing. I can even read the intricate tablets in the Sumerian language. I understand the enigmatic words in the stone inscriptions from the days before the Flood."*

Thus, contrary to what he had learnt in school, there had existed civilizations that ours, in some respects, cannot match.

Dave Long was born in Denver. His mother was of Italian descent. He owed his brown eyes to her, as well as his hair and olive skin. She died three weeks before his eighteenth birthday. After finishing high school, he attended the university in his hometown. Five years later he earned a Master's degree in political science. He successfully applied for a position at the office of Politics and People, where he is still employed.

There he met Rita Williams. After a three-year marriage, they separated, and subsequently divorced due to irreconcilable differences, as cited in the divorce filings. Together they had a five-year old daughter, custody of whom was granted to the mother. He visited little Lily as often as his busy job allowed. He took a month's leave of absence, following the death of his father.

Dave spent many days and nights reading notes, computer files and books, providing him with a clear insight in a world thousands of years ago. The *Sumerians* called themselves the black headed humans. They must have been a race of dark complexion with black frizzy hair, who established a dazzling civilization in the land of *Sumer*, hence the name by the surrounding tribes such as the *Akkadians*. He also read the epic of *Gilgamesh* and *Enkidu*. The English translation of *Enûma Elisj* recounts in flowery words the first days of an awakening humanity and its rich history thousands of years ago, now vanished in the shadows of time.

Surprised he wondered why he only now showed much interest in his father's great work. Previously, when his old man was still alive, he had occasionally showed moderate interest in these matters. He was overwhelmed with it and, to his great surprise, dragged to an ancient past. By acquiring more knowledge about this period in the history of mankind, he expected and hoped to bring clarity to the mysterious words his father uttered before he died.

He had developed a good relationship with his father's two assistants, who had no explanation for the nightmares they were still experiencing from time to time. They had recovered from the shock, but refused to take another look at the clay tablet, which he understood. He wondered in amazement how it was that all this was caused by studying a small clay tablet from ancient times.

Tomorrow he had another appointment with Carl and Don.

Dave Long had made himself a sacred pledge, to solve this mystery. What was puzzling to him was the fact that he too had seen the clay tablet, although he had not experienced those nightmares that had driven Carl, Don and his father to despair. Whatever did his father mean by "they are coming . . . they are coming," and damn, who were the ones who would be coming? Yawning loudly, he retired to the bedroom upstairs.

He woke up early the next morning. When he opened the newspaper, an article on the front page immediately caught his eye. Physicians had been inundated with complaints of patients from the small town of Siloam Springs, who suddenly were experiencing strange terrifying dreams. It was remarkable though, that especially people in the arts experienced dreams of a similar nature.

They were dreams of a strange world, where emaciated beings tried to penetrate the psyche of the dreamers. All the patients, with no exception, had the strong feeling that they had to fight, with all their might, not to be expelled from their body by these inexplicable demonic powers. Some of them did not want to divulge much, afraid to be ridiculed and declared insane.

Dave got excited, as his eyes swiftly went over the lines in the article. He read it twice and decided then to contact the newspaper to try and arrange a meeting with one of the physicians whom the patients had seen. He looked up the telephone number of the newspaper and dialed it on his mobile phone.

The doctor looked at him thoughtfully, after he had finished speaking.

"It's the first time I experienced this in my medical practice. It's a strange thing. I referred them to a psychiatrist for further treatment. Some of them did so reluctantly, but others categorically refused to do so. They did not think they were insane. A logical reaction of these people, by the way. So, your father had similar nightmares as my patients?"

"Yes," said Dave with a dreary face.

"How is the situation now with the patients?" he asked.

"I had to deal with six of these cases, including a very serious one."

"How so serious," Dave asked, looking tensely at the doctor.

"That patient is in bad shape. He is a science-fiction writer, who was severely delirious with fever before he fell into a coma." Dave was startled and jumped to his feet.

"Oh my God, those are the symptoms I observed with my dad."

For a moment it was dead silent.

He wondered anxiously what all that meant. Was the world on the threshold of something terrible?

A cosmic horror beyond the power of man?

This horror could not be confronted and combated with weapons. This was something of a completely different nature. It concerned the soul of man. This was the early beginning of the rise of an unknown threat engaged in a diabolical way to expel the souls of men from their bodies.

After a long period of silence, the doctor grabbed a pen, tore a sheet from a notepad and jotted down something. As he handed this to Dave, he said: "I also had a lengthy conversation with the grandfather of one of my patients, who is Native American. He gave a detailed explanation of the dream of his grandson, an upcoming painter.

"When he is in town, he always stays with his grandson. I think you should immediately contact this gentleman, who comes across as very wise and well versed in spiritual matters. Perhaps he will be able to explain these bizarre dreams further to you". After talking to the doctor for half an hour, Dave left, determined not to leave this mystery unsolved. Besides, his interest in this phenomenon was growing every day. This mystery was screaming for a solution. He planned to telephone the Indian gentleman for an appointment that very evening.

On the telephone, Matt Wilcox told him that his grandfather had already left for the Navajo Indian reservation, where he presently lives.

Matt invited Dave to his house for a visit. Dave took a taxi and went to see him. At an old house in a quiet street, he knocked on a weathered door, badly in need of paint. He was kindly invited inside. Dave sat down on a worn couch. On the wall an unfinished drawing of a beautiful Native American lady was attached with thumbtacks. While enjoying a cup of tea, Matt started talking. "I had a few awful dreams. After two grueling nights, I called my grandfather, who lives rather far from here. He empathized with me immediately and told me that he would come and visit as soon as he possibly could. Two days later he arrived by train and he telephoned me to pick him up at the train station. That night, after a few Indian rituals I did not understand, he asked me to go to sleep, which I did. That night, I experienced the worst nightmare ever. Early the next day, I woke up to tell him that his help, although appreciated, had been of no use at all. I found him in the kitchen, deeply in thought. He invited me to sit at the table. He had already made breakfast. With a serious look on his face, he told me what he had witnessed. He had not tried to wake me up. He said that I was delirious and spoke of something or someone I was wrestling with . . ."

"Do you still remember what that something or someone looked like," Dave interrupted. "Yes, sure . . . oh my God what a horror . . . it was a misty kind of appearance . . . scrawny. It was a woman. She looked very old . . . Her big eyes, at times, looked fiery and then again dull. Frightful!"

"And how did you wake up from that nightmare?"

"I struggled to wake up and finally succeeded. I woke up all sweaty and saw my grandfather standing at the side of my bed, holding a glass of water and looking at me, very worried. He said nothing and left the room. I didn't want to go to sleep anymore, but after tossing and turning for a while, I fell asleep again."

"Please tell me more about those uh . . . dreams."

"That's so weird. I dreamt that I was in another world, far from here. I saw that world in a flash and then that picture was gone again."

"Can you describe what you saw?" The tension in Dave's eyes was evident.

"Sure, I saw a hi-tech world like you see sometimes in science-fiction movies, really! That strange world appeared in my dreams twice."

"What else?"

"I don't remember. I think I had a dreamless sleep until the next morning."

"And what was the reaction of your grandfather, when you told him all this?"

"Well, I told you that he witnessed everything, sitting beside my bed."

"You dreamt of . . . *the Nakus* who once ruled over us and want to return," he said in a mysterious tone. Beyond that, he didn't want to share anything else, but he gave me an amulet for protection." Matt pointed to a small pouch made from rough leather with a similar thin black rope around his neck.

"He asked me to call him, when I saw him off. But before we went to the train station, he accompanied me to Dr. Will Jameson, with whom I had an appointment after the third nightmare. That's the doctor who referred you to me. Dr. Jameson spoke for a long time with my grandpa, who told me that he was of American Indian descent also hence, of course, their long conversation regarding the situation with the other patients, who had experienced those dreams too."

"That's strange, very strange. If I summarize all this, I reach the tentative conclusion that this is the beginning of a serious danger, which will plague mankind. What I don't understand yet, is that hypermodern world in your dreams. I am doing my very best to grasp all of this with my sober intellect."

"I would advise you to pay a visit to my grandfather. I have the feeling that he understands more. He would have stayed longer, but he had other urgent business with people on the reservation. He is a shaman with a rich and vast experience. Grandpa Jonathan is a remarkable man. He holds a degree in cultural anthropology from Arizona State University, and is specialized in the culture of extinct indigenous American people. Not only did he wander through South and Central America, but he also spent many months in Iraq and Egypt."

"But how did he become a shaman?"

"He had that since birth. He was born with certain gifts. His father was a great Navajo shaman. It's a tradition and inherent in my family. Every first son seems to be born with those special gifts."

"Interesting story, and how did he decide to go to college?" "He always had the desire to go to college, according to our family, that's why he moved to Metro Phoenix at a young age to live with relatives, in order to attend high school and later university.

Unlike others, he never forgot his *roots*. He devoted himself, after finishing his studies, to shamanism, to the delight of his father who died at a very old age."

After talking to Matt all afternoon, Dave went back home. He had stored the old man's number in his mobile phone.

"You should call him early in the morning. He is a very busy man."

Following a long and exhausting journey, Dave sat opposite the old Indian gentleman, who leaned on a beautiful thin cane and looked at him surprised and then serious. Wilcox kept staring, completely enthralled by him. There was briefly a strong emotion on the shaman's face. Dave wondered about that, but said nothing. The shaman cordially invited him to enter the cabin. They sat down on rough wooden benches. The shaman put his cane aside.

"This cane is more an attribute associated with the Indian shamans," he said with a broad smile. "I really don't need it to walk."

Jonathan Wilcox, who had passed the prime of his youth a long time ago and was now in his late sixties, spoke in trailing long sentences. He was a strong, tall man with light brown skin and a proud look in his dark slanted eyes. His face, with the distinctive American Indian features, looked serious. His long grey hair was tied back with a black ribbon. He had an aquiline nose, often seen in the images of ancient Native Americans. He reminded Dave of the great chief of the past century, *Geronimo*. At his father's house he had seen a picture of this great Apache chief of yesteryear. Gradually, the conversation revealed that this man was well versed in many subjects, which you would not suspect at first glance. He looked very simple, like an ordinary old Indian from some forgotten reservation.

Jonathan took out a typical Indian pipe with a long stem, carefully packed it with tobacco, lit it and took a deep puff.

The smoke curled up and had a pleasant aroma of a strange tobacco, Dave could not identify. The scent was quite peppery, he thought. When the smoke penetrated his nostrils, it gave him a spacey feeling.

Then the shaman began talking, with a serious look on his face.

"The greatest grace we received from the good God, to me, is the inability of the human mind to grasp and interconnect everything on earth. Our whole thinking is filled with ignorance of many things in this short human life on earth. And that's just as well. Hitherto science has harmed us little. The terrible views on the reality of the things around us, should they emerge from the darkness, will cause such shock to our conditioned mind that we might go nuts by these revelations and would want to escape to the peace and safety of earlier centuries, when we were made to believe in blessed fairy tales."

Dave listened to the old man with great interest.

"Such as . . . ?"

"Yes, such as religious teachings that are faint echoes of real events on this planet. Some of these teachings and their man made dogmas would end up on the scrap heap of our thinking, if we managed to get behind these truths."

"Could you give an example," Dave asked cautiously.

"The pure teachings of the old masters have been taken out of their original context and have been living lives of their own. An example of our distorted thinking: just look around you. The construction and actual purpose of the signs of the flame—we still cannot explain the pyramids of the ancient Egyptians and Mayas.

The theories regarding these mighty structures breathe the shortsighted thinking of many scientists. The ruins of *Baalbek*, the observatory at *Chichén Itzá* in Yucatan and the *Sphinx* must be laughing under the hot sun about these delusions.

Many shamans who still work among their people are a laughing stock, dismissed by many scholars of our times and before. We, especially westerners, dismiss them as silly superstition . . . and belief in ethereal enormities, spawned by primitive animistic thinking." The shaman paused and took a deep puff on his pipe.

Dave, who listened attentively, waited patiently for the answer and explanation of the terrifying dreams his father experienced before his death. He also waited tensely for the answer to the nightmares of Jonathan's grandson, the two assistants and the other victims.

"I appreciate your presence and above all, your willingness to listen to me," the old man said while getting up.

"Come with me." He motioned for Dave to follow him. Dave followed the old man and looked around curiously. It was his first time on a reservation. Before coming here, he had done some reading about it on the Internet.

The Navajo reservation in Utah in the western United States is home to the largest population of Navajo Indians, who often live in abject living conditions, poverty is rampant.

He clenched his fists angrily.—It's terrible, what the white man has done to these people throughout the ages—he lamented, shaking his head irritably as he continued walking.

After a short stroll, they arrived at a typical Indian tepee, in front of which stood an impressive totem pole. The artistically engraved animal figures—an eagle head, a thunderbird and three snakes were painted blue, white and red. "Welcome to my sanctuary," said the old Navajo laughingly.

"Wow! That's very impressive." Long pointed at the totem pole.

"They worked on it for a year and when it was completed, it made a great impression on me too," Jonathan said proudly. "The solemn inauguration with the necessary ceremonies lasted three days. Two old shamans on the reservation were assigned this task. Unfortunately, one of them died six months ago. George "Great Bison" turned 96 one week before his passing. That guy was full of humor. He was a very wise man, from whom I learnt an awful lot."

A wistful look appeared in Wilcox's eyes, as he said so. They took off their shoes and stepped inside. Jonathan turned on a lamp. On a dark brown bench Dave noticed a sophisticated, what looked like a very expensive laptop.

Before he could say anything, the shaman remarked nonchalantly: "Many years ago, I acquired a satellite dish." Laughingly he added: "You gotta go with the times, don't you agree?" Dave nodded and sat down on a wooden stool.

"I want to show you something interesting." He picked up an old suitcase off the floor and took out an object, wrapped in a thick brown cloth.

Carefully Jonathan placed a very old statuette on the table.

"Do you know what this object is?"

Dave shook his head, said nothing, squinted and studied the artifact. This ancient statuette was approximately 12 inches high and represented a woman with serpentine facial features. "*Nin-khursag, Lady of Life, the legendary goddess of the black headed humans, the Sumerians*"

Thoughtfully he looked at Jonathan.

"Where did you find this statuette?" The shaman looked at him mysteriously.

"This statuette was not excavated in Iraq, you know.

"If not there, then where?"

"In Guatemala."

Surprised Dave shook his head.

"Guatemala? That is a considerable distance from Iraq."

"Yes, indeed." Dave took the statuette carefully in his hands.

"Wow! Unbelievable, I hold here in my hands a statue that's thousands of years old. But how the hell did you get it?'

'I like to browse at flea markets. One morning I visited a small market not far from my hotel in Santa Cruz, where I attended a conference with representatives of indigenous peoples. We had a day off. I was there with no particular purpose, when I found myself in front of a stall with a display of pottery and other items. Then my eyes fell on a statue that seemed pretty filthy, with caked on mud. I purchased it from a little old woman behind the stall, mostly to give her some money. The woman looked very poor, but she was very friendly. When I arrived at my hotel, I carefully cleaned the statue. You can imagine my surprise, when I discovered that it represented a woman, totally different from the Mayan statues I had seen previously. To my bigger surprise, I saw an inscription in *Cuneiform* at the bottom of the statuette. Here, look." Jonathan pointed at the inscription with his finger. With narrowed eyes, Dave studied the text.

"Damn, that is indeed *Cuneiform* script."

"It says *Nin-khursag, Lady of Life*."

"But ahem, do you know *Cuneiform* too?"

"No," laughed the old man. "I took a picture of the caption and emailed a scanned copy to a fellow professor with knowledge of these old symbols. He did not ask me any prying questions."

"But this statuette is invaluable; museums would pay a lot of money for it"'

"Yes, indeed, but do you know what's so strange and mysterious about this?"

The shaman took a big puff on his pipe. The smoke curled around.

"This is so strange and incomprehensible. A few weeks later, I returned to Guatemala, on my way to Nicaragua for a follow-up conference. The market was not far from the hotel, as I previously told you. I was curious about the origin of the statuette and wanted to ask the old woman about it. But I saw no old woman behind the stall, but a strapping young man, who was selling other kinds of goods. When I asked about the old woman, he looked at me dumbfounded, and told me that he's had that stall for the past two years, leasing it for a small fee. I told him that a few weeks back, it was on a Friday, I found an old woman standing there, who was selling pottery and other ware. He looked at me smiling like I was crazy. After some thought, he said that on that particular Friday, he was absent, due to illness of his daughter. Perhaps someone else had used his stand, although he doubted it, as this had never happened before. I left flummoxed and disappointed, after questioning the other vendors about the old woman, I went back to my hotel. They all did see an old woman behind the stand that Friday, but she had left after about fifteen minutes. Strange, don't you think? But I learnt in this life there is more between heaven and earth, than we humans are able to understand."

"But what does this statuette have to do with the nightmares of your grandson and the others?"

"Patience young man, I will get back to that in a moment. Through my intensive study of secret rituals, I was able to cast a cursory glance at hidden centuries, the thought of which still makes me shudder. An average, uninitiated person would definitely end up in a madhouse, if he would go through what I experienced then." The shaman paused briefly and took a deep puff on his pipe. He looked very serious and continued. "When I associated some totally different data and went into a deep trance, I received in that trance a flash of truths, confirming all my suspicions.

The shaman exhaled, cleaned out his pipe in an ashtray, and packed it again with tobacco. Nervously Dave drummed with his fingers on the table and stopped when Jonathan started talking again.

"Your father, when he came out of that coma, told truths that integrate seamlessly with ancient writings. These dangers are beyond our comprehension and will come down on us, mercilessly."

"And how can we avoid or ward off and fight this evil? And what does this danger really entail?"

The shaman shook his head. 'One cannot possibly fight it, at least not with the current resources available to modern science. This has nothing to do with physical, material things.

The Incas and Mayas left behind impressive writings, in which they warned of this danger that would befall humankind. Unfortunately many of these precious texts were destroyed by the Spaniards, because they were not consistent with the prevailing Christian doctrine and beliefs. What remains, is a small number of fragments and citations of these great peoples. The sacred book of the Mayan civilization, the *Popolvuh,* now released in various reprints, only contains very little of this huge legacy.

The Mayas and also the Incas were deathly afraid of that what is happening to us at this time."

He took another short break and looked intently at Dave.

"They were terrified of the second coming of the gods!"

NINE

"Which gods, for goodness sake" Dave asked surprised. A scornful smile appeared on the old man's face. He took a deep puff. Angry clouds rose from his pipe. A grimace appeared on Jonathan's face.

"Gods! They were extraterrestrials with cutting-edge techniques, pulling mankind from the deep darkness in which it wandered hopelessly thousands of years ago."

"Excuse me, but I don't see the connection with the nightmares of those people, including your grandson and my father," Dave sighed wearily.

Jonathan motioned with his pipe.

"One of the most secret writings, a copy of which I have here, the same is clearly stated as in *Cuneiform* script on the clay tablet that was fatal to your father, after he examined it. But," the shaman paused briefly. He sucked on his pipe again. "The Mayas go much further and clarify more."

"They will chase away our own light and shine their light in our homes, tired and weakened as they are in their degenerated homes on the verge of collapsing."

"Enigmatic words I have been studying for years. Only now do I understand the extent of this mysterious text.

When you told me about the words your father uttered before his death, it began to dawn on me. Even more so, the nightmares that plagued him and the others, gave me a clear picture of the tragedy humanity is about to endure.

We will be expelled from our body and they will take possession of it!"

"Who will take possession of our body and why?"

"That is the crux of this problem, which we have to find out now."

"It is so complicated and hard to understand."

"Yes, difficult to understand, but it's not that complicated. Right now, I will not divulge much about my thinking . . . I suspect something . . . I . . ." He stopped talking abruptly and puffed nervously on his pipe. Dave did not respond, but sat quietly on his stool and thought—contact with the bizarre is often more frightening than inspiring. The sinister idea of what would happen to humanity and this civilization evoked in him an increasingly somber feeling.

When Dave remained silent, the shaman said: "I have a strong feeling that the answer to this dramatic event is hidden in this statue."

Dave was startled and looked at him surprised. "How . . . what do you mean?"

The old Navajo did not speak, picked up the statue and turned it upside down, studying it with a faraway look in his eyes. "I will go on a strict diet and into a deep trance and, hopefully, the truths will come to me; truths which, frankly, frighten me terribly."

"You? . . . Fear?"

"Yes, fear for what lies ahead. It's like playing in the woods with fire and dry leaves and knowing full well that can generate a huge fire, with all-consuming flames."

"Oh and when will you go into a trance?"

"Tomorrow I'll start fasting. I will take you to the bus stop now." He looked at his watch.

"You will be able to catch a bus in an hour and a half. We will stay in touch by telephone and email."

He took a business card from his wallet. "My email address is on it. Call or email me in a week. I hope to have better news for you then." They decided to leave. Both men, talking softly, walked to the bus stop an hour's walk farther.

TEN

The meeting with Jonathan Wilcox had left a deep impression on Dave. He hoped the old shaman would soon come up with a solution to the terrible evil that had befallen mankind. Since two days he had been at his father's house in Siloam Springs. It was time for the evening news. He settled comfortably in an easy chair. Dave began worrying terribly, when the newscaster reported two dramatic events in the early morning. One man committed suicide by jumping through a window, screaming fearfully. The other case involved another man, running through the streets in his pajamas, screaming loudly and clutching his head. A night watchman who, alerted by the screams, came to the victim's rescue witnessed how the man, before he shockingly expired, groaned that something or someone was trying to penetrate into his mind.

After the news broadcast, the television station was inundated with telephone calls of worried people, who had also suffered indescribable nightmares in their sleep. A very excited lady said that in her dream she tussled with something evil, which tried to take possession of her body. It was a mentally exhausting struggle, lasting many hours, before she was left alone in the early morning, completely exhausted.

"Damn! These incidents increase in number every day," he growled disturbed.

The next day the local newspaper reported a confused letter to the editor, in which a respected Catholic clergyman predicted a bitter future from visions he had several nights. Dave decided to immediately contact the priest, whose parish was not far from the street where his father used to live.

After many attempted telephone calls, he finally got father Jim Hanson on the line. The weak and shaky voice of the clergyman convinced him that things were not going well for Hanson. He made an appointment for the next day. That appointment never occurred, for the priest was admitted to a psychiatric ward. He died a few days later, after continuously screaming for help. The newspapers carried the story splashed across their front page and also reported on cases of panic attacks with six other people, who exhibited the same symptoms. They had all cried for help before being hospitalized completely insane. Students at a high school in town experienced the strangest incidents in the middle of the day. It looked a lot as if they were possessed by demons from hell, which had chosen them as prey. They rolled on the floor, punching every which way, as if they were fighting against invisible forces. After an hour or so it was over. There were more of such bizarre incidents at other schools in Siloam Springs. The next few days the news was totally dominated by reports of nightmares and insanity of the people in this small town. These incidents became world news. CNN journalists came out hungrily with their recording crews. The Arabic news channel Al Jazeera also showed great interest. The Washington Post devoted a full front page under the heading: *"The Siloam Springs Syndrome."*

She sat opposite him, her palms pressed against her ears. She was a beautiful white woman in her thirties, a tall slender body, extremely feminine. Gorgeous legs. Wavy brown hair. When he took a closer look, Dave noticed that her face was handsome rather than beautiful. She introduced herself as Ann Blackson. Her voice had a deep alto, completely befitting her personality.

"Ann, please tell me quietly what happened to you." She raised her head. Teary brown eyes stared at him. She pulled a tissue from her purse, blew her nose and haltingly began telling her story.

"It started last night when I saw them in my dream." She paused, twisting a strand of hair.

"Dream?"

She swallowed, clearly trying with all her might to overcome the impact of that terrible night.

"No! My God, it was a nightmare and so real, I can still see everything. First I heard a lot of voices talking past one another, strange sounds I could not make sense of. They were strange languages. Then I saw them . . . oh my God, such horrors, mangy emaciated beings with watery and feverish eyes, fighting to enter into my mind. Everywhere dark shadows overlapping one another . . . I resisted violently with everything in my power. I fought them and also fought to wake up. I could not. It was as if my consciousness was being sucked out. I had a strong feeling that I was losing my mind. All of a sudden the voices and faces faded. It seemed as if they were being driven away by another force. A face appeared in my consciousness. A strange, beautiful woman looked at me pityingly. She wore a long blue robe, her hair hung loose down over her shoulders. Strange sounds from her mouth reassured me. She smiled at me and stretched out her hands to me . . .

That's when I woke up."

"*Nin-khursag.*" The words came softly, almost in a whisper from his mouth.

"*Nin* . . . what?" she replied surprised.

"*The Lady of Life.*"

"You speak in riddles."

"Allow me to tell you something very interesting."

"I'm listening." She looked curious. He folded his hands and began talking.

In her eyes appeared alternately different expressions: surprise, shock, disbelief, and finally relief.

He got acquainted with Ann after a program on local radio about the strange events of recent times. Ann Blackson was an upcoming writer of short stories. He had called the station after the show and asked for the presenter to whom he explained who he was. He received an enthusiastic greeting from Dick Connelly, who had interviewed Dave's father once on a completely different subject. After a short conversation and an invitation to come by the studio sometime, Connelly had given him Ann's telephone number.

"And now you know what's happening. It sounds incredible, but it is the truth. How this macabre story plays out, is still shrouded in mystery."

"I would like to meet that friend of yours, sounds like a very interesting man and, furthermore, a real shaman. I have some interesting books at home about shamanism."

"You will definitely meet him. What happened to you is also a bright spot in all this. I am curious about his reaction."

"So am I," she said, a hesitant smile appeared on her lips. They continued talking a while longer. They exchanged personal contact information. In a much better mood Dave returned home.

A few days ago he seriously thought about destroying the clay tablet, but something kept him from doing so.—That would have been a huge mistake, he thought, as he impatiently waited for a telephone call or email from Jonathan.

He felt strongly that Ann would play a major role in this drama. He emailed her twice and also telephoned to tell her that he was waiting for an email from Jonathan. Two weeks had passed already and still no sign from the shaman.

One evening, as he was checking his email, he saw a new message: *"Dear Dave, please come to me as soon as possible and bring the clay tablet with you. I have discovered something very interesting. Call you later. Jonathan."*

ELEVEN

He had been waiting for the shaman for some time, but the latter never showed up. He rang, but received no answer. The device rang, but no one answered. A disturbing feeling slowly got hold of his whole being. He took a few minutes to think and decided to walk to the place where the tepee was located. He thought—I hope I don't get lost, because that would be terrible.

When he got off the bus, the hands on his watch stood at five thirty. It was now fifteen minutes past the hour and it started to get quite dark. During his first visit, he had not paid much attention to the road on his way to the bus stop, inasmuch as he and Jonathan were engaged in a lively conversation. That would now avenge.

After walking for fifteen minutes, he wondered desperately whether he was on the right track. The area was pitch-black. His feet, at times, sank in the loose sand of the road. Carefully he continued walking. The sky was darkened by black grey clouds with here and there a faint flicker of a star. There was rain in the air. "That too," he growled frustrated. Lugging the backpack containing the clay tablet was now very disappointing. He had neatly wrapped the clay tablet in old newspapers to prevent damage. You just never know what might happen. After another ten minutes or so he gave up. Tired, disappointed and also angry, he sat down on the withered grass on the side of the road.

Surely, something must have happened to Jonathan. He grabbed his mobile phone from his pocket and dialed the number. In vain he waited for the kind voice of his old friend. After a while he stood up again and disoriented he walked further. Suddenly he saw light in the distance.—My God, people . . . they will certainly help me find the right direction. Enthusiastically he walked further. After a few minutes he began to recognize the area. That big oak tree was not far from Jonathan's tepee . . . ? That must be it. Then he saw the contours of the cabin, which was not very far from the tepee. "Thank God, I'm here," he said relieved. He quickly walked towards the tepee and opened the tarp marking the entrance. He entered quickly and looked surprised at Jonathan, who sat on a white buffalo hide, his legs crossed in meditation posture. He saw at a glance that the Navajo was emaciated. He was still staring in surprise at the shaman, who opened his eyes slowly and looked at him sharply.

"Mr. Jonathan, how are you? I was worried about you and telephoned you several times. Are you still in trance? I apologize, but I hurried over to see you after your email message."

The shaman still looked at him silently and wonderingly. That stirred up bewilderment in Dave. After he carefully placed the backpack with its precious contents on the floor, he hesitantly moved closer.

Something was wrong!

He took a closer look at the Indian, who cocked his head to one side and stared at him, as if he were a stranger.

Suddenly there was that voice, strange and distorted. He spoke in a language Dave did not understand. Was that a Native American dialect? No, this language was completely different. He said something else. Strange unintelligible sounds came from his mouth . . . . "*Gnftaa . . . rtgfha . . . kxwng . . . ftyhfbsnsjhs . . . dfhbsh . . .*"

"Jonathan, what the hell are you saying? I don't understand anything."

The old man continued looking at him with his head cocked. Dave could not let go of that penetrating gaze. For a few seconds he stood there undecided, wondering what to do next.

All of a sudden Dave realized the truth.

He shook his head confused and frightened.

"My God, it cannot be what I'm thinking?"

Abruptly there was loud, hollow and persistent false laughter. The shaman's face all at once looked daunting and evil. Chills ran steadily down Dave's spine.

–Something really terrible is going on here!

He had to get away from this place as soon as possible.

He started leaving the tepee as fast as he possibly could, when he noticed the laughter was weakening and slowly faded away.

Silence reigned in the tepee now. Only the sound of his breathing broke that horrible dead silence.

What happened next, made his hair stand on end.

Suddenly a very slow icy blast began blowing through the tepee, followed by a hot breeze.

Surprised and fearful he stood there shaking on his legs. Baffled, he impulsively closed his eyes. At that moment he understood what was happening here.

Two forces were colliding in a devastating battle.

Mentally he saw a lot going on.

He opened his eyes again—still stunned by what was happening around him. The tepee began to move.

It shook violently back and forth. He wondered desperately whether the tepee would hold up . . .

"Soon everything will be blown away . . . ," he moaned hoarsely.

His gaze fell on the shaman, who seemed to be engaged in a tussle against something indescribable. First he was hitting himself, and then he stretched out his hands as if he were strangling something or someone. He was throwing punches left and right, like a professional boxer. "I wouldn't want to be in the place of this invisible opponent" Long thought, as he anxiously followed the fight.

There was more . . .

The shaman fell to the ground with a thud and started rolling around. In the turmoil, the benches and the table were violently thrown to one side. Dave jumped terrified to the side.

After a minute, which seemed like an eternity, Jonathan stood up, breathing rapidly and staring proudly in front of him.

The cold air stream disappeared and the tepee was flooded by a pleasant breeze. In the tepee there was now a peaceful, unfathomable sphere, which completely overwhelmed Dave.

Jonathan took a seat on the floor, his palms pressed together. He muttered and seemed to be praying in Navajo. After a while he motioned Dave, who had remained standing, to also take a seat on the floor. Still baffled by what he had witnessed, Dave sat down across from the old man. Jonathan cleared his throat, looked up and then faced Long.

"I have much to tell you."

"Good heavens. Wow! Those were terrible moments just now.

What the hell is going on here?"

His voice sounded nervous.

"Yes, indeed. You witnessed what one could call a personality change. One moment he was inside me and then I was outside again."

"How? What? What was inside you and what were you doing outside?

I don't understand any of this." Dave shook his head. The shaman pretended not to hear. He beckoned Dave to pick up his pipe that lay in the corner. After Dave handed him the pipe, he rubbed it clean on his shirt, and slowly began filling it with tobacco from a pouch he had hanging around his waist. He pressed the tobacco in the pipe. As he took a deep puff, a serious look appeared in his dark eyes.

–Right now, he really looks like one of those proud Indians of yesteryear who smoked the peace pipe, Dave thought with a smile. Suddenly, the voice of the old shaman said: "I saw and experienced terrible things. I went to a world far from here and, in fact, close by."

"Explain please. I still don't understand any of this."

"I'll start at the beginning. After you left here days ago, I did some deep thinking before I went into trance. Before that, I also performed certain age-old rituals I learnt from my father through traditions of the ancient Navajo people. An indefinable impulse led me to carry out specific deeds I only now begin to understand better. Then the urge came again today to go into a deep trance.

During this trance, for which I applied certain old techniques, I began, within a few minutes, traveling with great speed through a vast space of deep inky darkness. My eyes saw nothing. It was an unbelievable sensation of moving forward, associated with thick, almost liquid-like air raging past me. After a while, the darkness disappeared and it became real light around me. I was floating into another dimension, where I saw things that are beyond words. I flew into a world which in terms of nature, seas, rivers, lakes, mountains and forests is equal to ours here on earth . . . and yet different. On my flight in this disembodied state, I flew across impassable deserts and white sands. I saw once powerful cities in decline. Encroaching bright green creepers, rare flowers, mosses and strange multi colored large mushrooms were everywhere. Those cities looked totally different than the ones here on earth—modern style cubist buildings, for my earthly eyes strange and unnatural cyclopean forms of an eerie non-geometric architecture.

But the biggest fear was yet to come."

Jonathan paused for a moment and bit the stem of his pipe, then started talking again.

"I was terrified when I saw them. Phantoms clad in dark robes, stumbling and moving about helplessly."

"What did these creatures look like?" Long asked.

"I was curious too and floated closer to take a look. They looked like us, only a lot taller in stature and with slight serpentine facial features. What I immediately and shockingly noticed, was their emaciated faces and the almost translucent skin on their arms and hands, deteriorated at the end of their long, now distorted existence.

This race should have long been extinct. But thanks to their, for us impossible to understand, advanced techniques, they managed to exist until now.

I began to grasp everything better!

What I saw, was the end of a superior great race, fighting in vain against its physical demise. Feverish bulging eyes staring at me, thin arms trying to grab me . . . eyes so big and vicious, flickering maliciously—you would certainly die of fear if you saw that. They attempted to oust my *psyche* and enter my corporeal body that I had left on earth. If that had happened, my return might have been impossible. But the daredevil who I am, and also driven by that strange compulsion forcing me to undergo all this with a specific purpose, which at that time was still hazy, I went on the offensive! The shaman took another puff on his pipe and continued his amazing tale.

"I began a spiritual battle with this alien race and forced myself into one of those bodies which, of course, was extremely dangerous.

Then a truth was revealed to me, of which I had a vague suspicion, weeks ago on earth, when you approached me about the nightmares and told me about your father's final words. I immediately understood the nature and the condition of these creatures. The flashing images

I received from the memory of my "victim" provided me with a deep understanding of the rise and fall of a truly great race.

Names of great personalities, in hieroglyphs, passed in front of my spiritual eye, which to my surprise, I was able to read also. In short, I began to understand everything much better. I visited in that trance state, the repository in a large dirty white building, where a lot of knowledge was stored in equipment not receptive to humans. Equipment, a strange kind of computers, run by thought impulses. I was also able to establish, by tapping into my "victim's" memory, that there was no trace of any religious life whatsoever with these creatures."

"Not a single trace?"

"Nothing whatsoever.

They have no temples, churches or mosques. There are no images anywhere or places for holding religious services.

They don't know priests, preachers or clergy. The existence of such institutes is completely unknown to them.

Apparently these beings do not have the need to worship anything."

"Do you think that in a distant past they might have had some sort of religious thought?"

"That is possible, but apparently, after thousands or millions of years of evolution, they had outgrown this thinking. But they certainly had knowledge of an immaterial world; therein they were quite advanced".

"Hence, they were very spiritually attuned."

"Yet many chose the path of evil . . . the obscure, otherwise they wouldn't want to destroy us in this way," Long said indignantly.

"Evil thus! I got the distinct impression that they do not want to give up physical existence and are desperately clinging to tangible objects."

"That's what I believe too." Dave grumbled.

"Not being able to distance themselves from their grand and glorious past is the main reason for their actions," the shaman said with a somber look in his eyes.

"And what happened to you next?"

"After gaining the knowledge, I exited that creature's body and wanted to return to my corporeal body here on earth. I realized that would create huge problems. One of them had successfully entered my physical body. I had to engage in a terrible fight to expel that damn thing from my body. You follow, yes?"

Jonathan stared at him intently.

Dave nodded, trying very hard to absorb everything he was hearing. There was astonishment and dismay in his face.

"So what I just witnessed . . . the scuffle you were in just now . . . ?"

The old Navajo nodded.

"Now I begin to understand. I also learnt why I was chosen and guided to do things to combat this disaster that will befall humanity in the coming days."

After a few seconds of silence, Jonathan said: "The events we have been dealing with the past few days are only the beginning."

He sighed deeply and took another deep puff on his pipe. The smoke curled in circles upwards. With a tense gaze Dave followed the smoke, but his thoughts stuck with the almost incredible events Jonathan had experienced.

He was startled when the Navajo began talking again. The Mayas and other ancient civilizations indicated that Earth, in our time, will have completed a 26,000-year cycle around the sun. During the completion of this cycle and at the beginning of a new one, our astronomers will observe a new planet. If we summarize all this, I reach the following conclusion: the appearance of this planet will have disastrous consequences for us *earthlings*.

"You and I are the only ones with this knowledge! What we have ascertained, is the desperate struggle of that race to survive. It is those beings, who at the end of their physical existence have opened an assault on us *earthlings*. This knowledge is a heavy burden to bear alone. The people around us will not understand."

The shaman let out a deep sigh. His gaze shifted to the statuette on the table.

"You also told me that there is something hidden in this ancient statue," said Dave with an undercurrent of tension in his voice.

Jonathan nodded and briefly closed his eyes.

"The statue of *Nin-khursag* contains an answer or a clue as to how to proceed. It was no coincidence that on that day at that exact moment I found myself on that market.

"It was destined to happen!"

"She, Nin-khursag, her not-reincarnated soul is still fighting, after all these thousands of years, against the other one . . . Enlil."

"I don't get it." Dave stared at Jonathan with a blank expression.

"This battle which is conducted on a spiritual level, has now reached an unprecedented climax.

How this will end is up to us, yes you and me.

We are both facing extremely hard spiritual work.

And you will also play a role in all of this."

"I will . . . . ?" Dave asked surprised.

"Haven't you noticed that you are not susceptible to the penetration of these beings into your body? There is, I suspect, a protective shield around you. From the first day I met you, I saw at first glance a magnificent orange colored aura surrounding you."

"And I still remember your surprised look when we first met several weeks ago, said Dave.

Oh Jonathan, there's one more thing."

The shaman looked at him questioningly, while Dave cleared his throat.

"I had an interesting meeting with a young woman, who experienced the following in her dream."

Jonathan listened attentively, without interrupting him. When he finished, the shaman reacted enthusiastically. "This is great! This woman also has something that protects her, and on top of that *Nin-khursag*." He spoke her name reverently.

"That gives us hope. The spirit of the primordial mother will certainly support us in what we are about to undertake. I have a good feeling about this," said the shaman.

"So do I. But what I don't understand is the fact that my father and his two assistants, after studying the characters on the clay tablets, were affected by those terrible nightmares, which ultimately led to my father's demise."

The shaman looked at him thoughtfully.

"Memories remain stuck everywhere like rust stains. You can find them in many old buildings. But aside from memories, curses cast a long time ago in association with dark rituals, also remain stuck to objects and perhaps those curses were activated when they were studying the text. I don't have another explanation for this." Jonathan shrugged briefly, as if shaking off a burden. He knocked the ashes out of his pipe and stretched, muscle for muscle.

"Let's take another careful look at the statue." As he said so, the shaman got up and carefully took the statue out of the suitcase. For a few long minutes he stared with narrowed eyes at the image of *Nin-khursag*. Then he said softly: "It's beautiful." He bit on the stem of his pipe, turned the statue upside down and shook his head.

"Maybe there is something in *Cuneiform* inscribed somewhere," Dave suggested carefully. "Hey, that's a thought."

"I have an idea. We could take this statue to a lab and have it looked at under a digital microscope. You never know."

"Do you know where we might find one?"

"The university where my father lectured has one. A few months ago they substituted the old microscope for a very sophisticated one, my father had told me enthusiastically on the phone one day, when I asked how his work was going.

"Please contact them tomorrow morning. We have no time to lose." Both men talked for a long time then decided to call it a night. They returned to the cabin, which had sleeping accommodations for three.

TWELVE

Jonathan looked up at the silver grey morning sky. He stood in front of the beautiful building of the Bentonville hotel in Siloam Springs. He searched his mind—would the microscope in the lab reveal what he and Jonathan were not able to see on the statuette with the naked eye?"

They had an appointment at the lab with Dr. William Stone later that morning. They arrived last night. Dave had emailed from the reservation to the university, followed by a telephone call. William had been a good friend of his father's and the appointment was arranged quickly. Hopefully the examination would have an impact. He also telephoned Ann to find out whether she was free. She answered in the affirmative and promised to be at the university on time.

An hour later they were all at the lab waiting anxiously for things to come. With a mesmerizing look in her eyes, Ann held the statuette in her hands. "Wise *Nin* thank you for your protection," she whispered softly. "Yes, she was the one in my dream. Look at the robe on her and that beautiful face. The maker of this statue has delivered a masterpiece, indeed." She felt a slight shiver run down her spine. She handed the statue back to Dave, who in turn, handed it to Jack, William Stone's assistant. He looked at the statue from all sides while whistling through his teeth.

"Wow! This is a really old, beautiful specimen."

They were able to follow everything on a big monitor. Parts of the statue were projected large on the screen.

Suddenly Jonathan's excited voice called: "Stop, go back!" The assistant followed the shaman's instruction. They saw something very small on *Nin's* robe.

"Take a picture!" sounded the excited voice of the Indian again. After having taken a few pictures, they packed up the statue. With the prints neatly stored, they returned to Dave's parental home, where immediately on arrival, they began studying the pictures on which something was engraved in *Cuneiform*. Without delay, Dave sent a scanned copy of the picture to a specialist in this script. After fifteen minutes the answer came by email.—*Search for Enlil in the depth of your consciousness and destroy him.* Surprised the two men and woman looked at one another.

"God, what does that mean?" Dave said aloud. Jonathan read the text one more time and then looked at that part of the picture where the original text was.

Silence entered the room. The shaman was thinking.

After a while he whispered: " . . . The depth of your consciousness . . . That uh . . . has to do with . . . He stopped abruptly.

"What do you mean?" Dave whispered with a tense voice.

Jonathan started mumbling . . . meditation? Trance? Descend into my consciousness? Dave gave the old man a blank look.

"We need to return to the reservation!"

"When?"

"Now, immediately! Every minute is important. Didn't you see in the news what's going on and didn't you see what's in the papers today?" Indeed, it was shocking news. The number of nightmare victims had drastically increased, also outside this small town. The world community of physicians, psychologists and psychiatrists, in short the entire medical world was upside down and could not find an explanation for this terrible phenomenon.

Another short article reported that astronomers had discovered an unknown planet that appeared suddenly next to the big Neptune. Dave brought this article to the shaman's attention. "Those two circles on the clay tablet with that mysterious line connecting them."

Jonathan uttered a muffled cry in Navajo.

"That's *Nibiru,* the planet of which the Ancients and theosophists reported in ancient writings! Holy Manitou, this means the end. If this planet emerges, the mental leap of its weakened inhabitants to the bodies of the people on Earth is only a matter of time.

They also wrote that this planet would be closest to Earth around this point in time." A wistful expression appeared on Ann's face.

"How did they get that wisdom and why, in heaven's name, should it occur now?"

The shaman shrugged.

"By cosmic laws unknown to us humans!"

They were back in the tepee on the Navajo reservation. The shaman had taken the necessary precautions to go into trance. Hopefully in that state, he would find an answer to the tiny *Cuneiform* characters engraved on the statue. Ann, much to her regret, had previous engagements and could not accompany them. They agreed to keep her abreast of further developments.

On his way to the reservation, Dave had picked up three different newspapers. With little interest he had read the headlines. He always does so, before reading the entire report or article that catches his attention. Suddenly he noticed a small article, which claimed his full attention. After reading it, Dave circled the article with his pen and handed the newspaper to Jonathan. "What do you think of this article?" he said worried.

A professor at the university, who had been suffering from nightmares, woke up one morning fresh and lively. He was freed from them, but according to his wife and two sons, he had changed into a completely different person than the man they had known. To their astonishment, he had immediately stopped smoking, a habit of his since he was eighteen years old—he was now a man in his fifties. The wife did not recognize her husband anymore. He walked and talked differently. He used to be a cheerful man, but since that morning he seemed much more serious and cautious. At times, he would stare at her curiously like a complete stranger and he was totally not affectionate anymore. His dog constantly growled at him, many times sneaking away with his tail between his legs.

The shaman quickly scanned the lines with his eyes.

"Do you know what I'm thinking and you undoubtedly are too?" Dave nodded, but said nothing. He nervously drummed with his fingers on the table.

"This article confirms, I think, the uneasy feeling I have been walking around with the past few days. Suppose some of these *Nibiruans,* so to speak, have succeeded to safely enter their new corporeal home—an earthly human body thus, we may fear the worst for the near future."

"This is only one of who knows how many cases which have taken place in the past few days."

"That will not surprise me, really."

"The journalist is not privy to what we know. Of course, he doesn't know that the nightmares were caused by the physically weakened inhabitants of a planet, who desperately need our bodies to survive."

"Hence, if we may believe this story, this is a case of a successful soul transformation. And what happens to the soul of the university guy?"

"He either wakes up in the body of that weakened creature that made the successful soul transformation, or he will forever wander about as a lost ghost," the shaman said with a dark face.

"*Enlil* must be found. He represents the evil that is coming at us. You're going to play a primary role in this." The shaman did not pay attention to the surprised look appearing in Dave's eyes. He took an old parchment star map from a drawer of the table. He studied the map and subsequently laid it on the table.

"Let's continue this conversation later. Now there's something else. In a few minutes I will go into a trance. What you need to do is the following. When you see the fingers of my right hand move, you need to dab my face with lukewarm water from the jug over there until I regain consciousness."

"Okay," Dave whispered, tense and utterly confused.

He tried to make sense of the words spoken by the old Indian. He was tense and nervous.

This would be his first experience with something like this.

"So, please pay attention to the fingers of my right hand, only when they move, will you do what I just told you. You will use the water from that jug." The shaman pointed to a deep brown earthenware jug on the table. "Don't be afraid or surprised by what you're about to experience. I will be going from a trance to a voluntary suspended animation. My body will remain in this physical world, but my *psyche* is going on a journey, where all boundaries fade and distances no longer count.

The shaman sat down on the white buffalo hide, his eyes closed and legs crossed in yoga pose. He mumbled a few words in Navajo . . . stopped . . . mumbled a few more words which transitioned into sounds, then he sank into a trance taking him far away . . .

Dave looked at his watch. He was worried. Three hours had passed and still no sign from the shaman, indicating that he was back in his body. His eyes were transfixed on the right hand of the old Indian. Suddenly he saw the fingers move slowly. "Thank God, he's back," Dave whispered relieved and nervous. He took a dedicated cloth and started dabbing the old man's face. To his surprise the water in the jug was still lukewarm.

After ten minutes, Jonathan slowly opened his eyes and sighed. A hesitant smile appeared on his lips that quickly disappeared. He looked serious. Dave, who did not know what to do, looked at him helplessly.

"I went far away, very far away."

"To that world over there?" Dave gestured upwards.

"Uh . . . you mean the planet *Nibiru?*"

"No, I did not go there. We came together."

"I beg your pardon, but I don't follow . . . came together? Who came together?"

A brief silence ensued.

"Let me explain to you young man. There are other shamans around the world. You can find them in Canada, South America, Asia, Australia, Africa, and everywhere you go. I met two other shamans, who are deeply concerned about the coming events, which they were able to see in transcendent state. We met in our spiritual body at a place in distant India. Naturally, this may sound incredible to some, but you experienced a thing or two in this tepee, so for you it's certainly no mystery. And with me you will have more of this kind of experiences.

Long said nothing and thought "these deep occult happenings frighten me terribly sometimes, but they are also incredibly fascinating."

Jonathan stood up, sighing deeply.

"I'll be back in a few minutes."

With a deep frown on his forehead he went outside.

Dave pondered the previous events: the death of his father; the tragic nightmares. All these incidents welled up inside him like a horrible flood.

"No," he whispered. "No . . . No . . . No . . ." The flood inside him kept rising.

"How many people on earth are aware of this danger? Very few, at the most he, Jonathan and two other shamans. How would one convince other people of the sinister things that were coming? Who would believe it?"

Anxiously he awaited the shaman's return. What the hell was he doing outside? He heard footsteps. The sound came closer. The shaman entered the tepee and sat down. He looked serious and scratched his chin. Then he began talking again.

"The three of us have joint our spiritual forces. But that is not powerful enough to avert the danger and save humanity."

"And what did you achieve with respect to that phrase in *Cuneiform* on the statue?"

"Nothing yet, but that's only a matter of time. The text sounds mysterious, but it is not. Human consciousness is very comprehensive. In essence, human consciousness is the brain wrapped in a material body, a typical anthropoid feature. But it is also able to exit the body and become one with that represented in everything. In deep trance and meditative state, it realizes that it has become one with the vast sea of *"All."* This is a natural and graphic description thereof." He paused a few seconds and looked straight at Dave.

"I went outside just now to look at the night sky. The stars floating along the ink-black robe of night, brought confirmation of certain ancient texts."

"But what's so important about those stars? And confirmation of what?"

"I had to convince myself that they were in the correct position relative to the newly discovered planet by our astronomers that is much older than Earth."

He pointed to the old map on which stars were depicted. With a compass he made a circle around the mighty Neptune and other stars. He whistled excitedly through his teeth. He pointed his finger at a star pictured near Neptune.

"Do you see that?" Dave nodded. "That is clearly a star, but what's so strange about it?"

An exciting glow appeared in Jonathan's brown eyes.

"That is . . . *Nibiru*!" Dave looked more carefully at the old parchment.

Suddenly the shaman said aloud: "Everything is just right!"

He looked sharply at Dave.

"And you my friend, you are here at the right time and moment!" Dave did not understand and looked surprised at the shaman.

"Dave this is a very special moment. He looked at him with an inscrutable expression in his eyes. Now it's your turn to go into trance and figure out the truth of things that still remain in the dark.

"Who me?" Dave asked surprised.

"Yes . . . you!"

The shaman walked to the rear and grabbed a few attributes, including a flute and a small drum decorated with eagle feathers.

"The future may also change the past, provided the right doors are opened to do so. You are the designated person for this purpose, dear Dave.

Now, do you understand the text?"

Sink into your consciousness and find Enlil . . . you must destroy him. One god against another god—fighting in non-material form to the final victory . . . or defeat. It all depends on who is the strongest and who is the weakest."

"Must I go into trance also?" Dave looked at the shaman in disbelief.

"An important requirement before you open up yourself to the trance state, is cleansing of your body and soul." His breath escaped his mouth with a hissing sound. The shaman closed his eyes for a few seconds.

Suddenly it sounded: "Fasting!"

It was momentarily silent. We have no time to lose, though. You would actually have to fast for a week or two. But in this emergency we cannot lose one day—

we have to drastically reduce that. Prepare yourself, for it will be three days . . ."

THIRTEEN

The shaman's monotonous chanting and the drumbeats slowly faded in the distance.

Dave sank into the trance . . .

The sharp light disappeared . . .

Slowly the darkness came to him, safe and protective with a glow swirling around him in progressively smaller circles. Pulsating vibrations permeated his body.

He felt the weight of his body falling off him like a heavy cloak. He changed into pure energy and had the feeling that he was rising slowly-forsaking gravity gave him an unbelievably warm sensation. He kept on rising to dizzying heights. His mind's eye glanced down.

He flew low over an inky sea extending to the horizon, where sunlight unsuccessfully tried to break through grey clouds in alternating malicious compact forms. A terrible fear came over him as he flew increasingly fast upwards. Sounds from afar softly penetrated through him. It was like the buzzing of thousands of bees. They became louder, then softer, louder again, then again softer. The gloomy greyness around him was occasionally illuminated by flashes of bright light, getting brighter and brighter.

Suddenly it was very light. The darkness was shattered as by a hard hammer blow. In his mind now swirling colors passed by: red, light purple and red again, slowly dissolving in a vortex of green and white . . .

Pure consciousness!

He was free!

His consciousness hurried past magnificent landscapes, blue lakes, and high mountains with snow-covered peaks. His *psyche* flew over green pastures and forests. Unprecedented colorful flowers in breathtaking colors flew past.

Then his *psyche* stopped for a moment.

Majestically regal she stood there on the flowerbed . . . she who was responsible for the creation, the distant ancestor whence he originated. Never before had he seen such a beautiful and divine appearance. Her light brown complexion was glowing and her almond-shaped eyes were captivating. Her narrow high forehead was framed by gorgeous long hair that fell to her waist and a long robe that reached to her feet. Eyes looking at him worriedly. Sounds came floating from her full lips and found their destination in him.

The sounds were a symphony within which thousands of faces whirled around. The music filled the air and thundered around him, grabbing him and immersing him.

He became one with this symphony.

His *psyche* was dragged into a downward flow of memories of past centuries' events . . . Thousands of lives were floating by in many varied incarnations. He was dragged along further and further, until the beginning of the genesis of man—where it all began.

He was enriched with knowledge of the how and why of cosmic revelations, which in secret had been waiting for his arrival and awakening. He was immersed in truths, disconnected from the conditioned mind, during his last incarnation on earth.

His consciousness, shocked by the views to which it was exposed, threatened to fall to pieces.

The feeling that his consciousness would be shattered intensified, but an unknown force, which unbeknownst to him was coming from himself, kept his thinking intact and prevented him to forever getting lost, circling in that dark eternity. He became aware of a world full of sounds.

He flew towards a symphony awaiting completion, for which task he was chosen from all the billions of *earthlings*.

In his mind he saw a smile appearing on the face of the great *Nin-khursag* and he realized then that he was on the right track to reversing the danger to which mankind had been exposed.

He slowly floated back down and merged with his body.

Dave Long felt his body breathe slowly . . . in and out. The regular beating of his heart caused a glowing bliss soar to his head. He slowly opened his eyes and looked into the smiling face of the shaman.

"Welcome back and how do you feel now after all the experience gained in those worlds?"

"Tired . . . and run down," Long answered. Physically he felt fine, but his spirit was in a mood that could be best described as empty and distant. The shaman looked anxiously at Dave. He noticed how clearly weakened the young white man was, mentally drained by the deep trance.

FOURTEEN

Their cities, designed and built with superior intelligence, were slowly declining. The race was in the final phase of its glorious existence. Physically they continued to deteriorate, despite the high degree of development in understanding and manipulating their physical condition. They injected their bodies full of powerful substances, nutritional medicines and other stimulants and lived on the top of their consciousness. The overall weakening of their bodies, which began annums ago, was unavoidable in spite of all the feats of these aids. Just like a plant withers after a spectacular flowering period, so too could one describe the decline of this race. Civilizations are able to renew in a cyclic manner, which in the past had also been the case with this species, but this renewal also has an end phase of rigidity and absolute standstill.

The race, through a long biological evolution and the use of certain techniques, had gone so far in an upward direction that nothing interesting remained to be explored for their thinking.

Underlying conflicts and wars had long been overcome and the memories of these barbaric times could only be found in chronicles stored in sophisticated computers. Once material matters were no longer of interest to them, they had plunged with enormous interest into the thorough study of the mind. They advanced very far with their experiments in this field, resulting in methods to possibly get out of this hopeless situation in which the species found itself. A detailed ingenious plan was waiting only for the auspicious time to be executed.

To save their consciousness, it was necessary to leave for good these decrepit and languishing bodies and reincarnate in new healthy ones, which could only be found on this small blue planet . . .

The serious faces of the old men in the white cool sterile area of the cabin were marked by wisdom and deep insights. Slow in his movement, it took *Lilta* a few seconds to turn his head in the direction of the big screen, where the calculation was projected in images. The time had come for *Nibiru* to emerge from the darkness of the universe for its closest approach to the blue planet. After thirty thousand *annums* this was again the case. From ancient chronicles they were able to distill that *they* were the ones responsible for the fact that the barbaric race inhabiting that planet then, was able to see the *light*. The *light* that would enable them to develop at a level equal to that of their creators'. Since their departure from that planet, they had shown no more interest in these *earthlings*. The interest only came after their physical condition began to deteriorate deplorably and they had studied all possibilities to get out of this miserable situation. This interest also arose when a theory was developed to transfer their consciousness to beings that had physical similarities with them.

The ancient chronicles also indicated that specific inhabitants of that small planet, in their genes, showed remnant traces of their creators. That was a great advantage to making the developed theory a success, when it would be applied in practice. In vain they had abducted some of the *earthlings* and attempted to inseminate them with the seed of their race. But that was a big failure, for the simple fact that the seeds of their species exhibited an unprecedented degeneration and proved to be infertile.

They had concluded, after having followed these *earthlings* for many *annums* that they had become a creative, arrogant and cruel race with still strong animalistic tendencies.

Lilta felt tired as he sat down. This fatigue had become characteristic of his kind. They still grew very old, but physical deterioration caused a lot of discomfort. A long time ago they had to forgo their taste buds due to the downward spiral of their evolution. They kept themselves alive through intravenous nutrition. Propagation through highly sophisticated clinical methods proved to be a big failure every time. The genitals of both men and women of this race had atrophied to "dead" parts of the body.

The race, during hundreds of *annums*, had dwindled to a few tens of thousands. With melancholy and pain *Lilta* pondered the glorious past of his race. If nothing positive was to happen to them, only the chronicles and cities would remain as silent witnesses of this great civilization.

Lilta was startled when *Kadir* came stumbling in, panting and seeking support from a shiny metal doll. As he spoke, his shoulders relaxed somewhat. "If we don't hurry and are not able to continue the dance of life, it will be the end of our physical presence in this universe. On three of our colonized planets, only the ruins of once mighty cities bear witness to our presence in the past."

A tired look appeared on *Lilta*'s face.

"We are at a critical stage. If the theory, which the wise *Inanna* left behind in the memory banks fails in practice, that will be the end of our existence."

"So, tomorrow night you will try to bring into practice the much talked about theory." *Lilta* looked tired. His glance was not what it used to be.

"We'll begin with the seven selected candidates. I expect them here tomorrow morning."

"I have already activated the *thought units* for the transfer." He exhaled deeply. His voice sounded scared, yet he spoke with an undertone of joy. "I've lived in this mortal shell for many *annums*—untold hundreds of *annums* more than many others. I have the feeling that my body is a dry shell just like the skin of a pupa that turns into a butterfly, but first must be shed before that happens. He chuckled loudly. "Yes, like butterflies many of us will be flying in the new environment."

"Will you and I make it to be reborn like butterflies in the bodies of those beings on that small planet?" *Kadir* whispered anxiously.

Lilta stared at him with a tired expression on his face.

"We might not, but others will. We are too weak to accomplish that . . . but who knows."

They were silent for a few seconds.

"Let's not get ahead of ourselves, tomorrow we shall know the answer," replied Lilta suddenly in a somber tone.

Krita lay on his back in the machine. *Lilta* secured a soft black hood over his ears, just above the eyes. Under the hood, the small adhesive pads attached to his temples and forehead. *Lilta* waited a few more minutes. The tension in the cabin was palpable. The scientist closed his eyes and with thought impulses he put the unit into operation to make the transition to a chosen *earthling*.

The body of the first one selected for this process began to vibrate. Vibrations turned into shocks, accompanied by light moaning. Then, the emaciated body lay still. On the monitor appeared a soft pink color that turned into dark grey. The round green dots, indicating whether the heart was still beating, were completely gone.

"And . . ." asked *Kadir*, who had become restless.

"It was a failure, he's dead," *Lilta* responded disappointed in a flat voice. A grim expression appeared on his tired face.

"I'm going to try again. I will do so until we succeed."

"Wait a minute, shouldn't you check the transmitter to see whether it works, maybe it does not function as desired." *Kadir* fumbled nervously with his white coat that hung like an oversized sac over his scrawny body. "Just look at the screen," motioned *Lilta*, as he pointed his thin finger to a big screen, where the interior of the device was visible.

"It works perfectly fine. The problem is at the other end on the blue planet, where the *earthling* is dreaming on his sleep sofa."

"So, the *earthling* has offered fierce and successful resistance, hence the failure?"

"That's right." In a few minutes, please bring in the next one selected."

"First this physical body should be transported to the dissolution units.

Let's hope that the telecommunications carrier still functions."

He sighed.

"Everything needs replacing—not just our bodies, but also much equipment is failing."

After the fifth attempt the process succeeded. A hesitant smile appeared on *Lilta*'s tired face." Let's hope that we will also succeed with the next selected candidate."

"They have to pave the road to our success."

Both men took time for a short break, before calling in the next selected candidate.

"There is a big problem."

"What's the problem? *Lilta* asked preoccupied, while setting up a monitor.

"Those out there." *Kadir* pointed at the armored glass with a view of the green cliffs of *Dirta*, where the glow of a rising sun announced the break of day.

"They are becoming impatient and resentful. We have promised so much and they feel that it is high time we explained the next phase."

"But isn't that what we are doing right now," an irritated *Lilta* growled.

"We had to keep the final preparations a secret. They believe that we aren't doing anything, and many who have become impatient, have tried to make the transition through psychological means, in their dreams. And from what I have been told, some followers of our opponent *Hrsma* have managed to make the transition successfully. But for even greater numbers of them, this ended in a total disaster. Yet, they continue with deep meditation techniques and dreams only known to his followers, who are still clinging to the doctrine of the legendary *Nin*."

Lilta looked at *Kadir* with a grave expression on his face.

"I foresee a fierce battle on that planet of our own species among ourselves and the earthlings against all of us."

FIFTEEN

Dave Long and Jonathan Wilcox were talking softly, while enjoying a cup of hot tea. "It will be a difficult and emotional struggle. It's fighting against what could be best described as an abstract evil." They stopped talking briefly.

"But there are several thousands of them against us billions of people," Dave said, breaking the silence.

"Thus we are in a numerical majority, but what if they manage to occupy strategic positions? Suppose, for instance, they occupy the bodies of presidents, influential politicians, bank presidents, leading scholars, you name it." The shaman, while saying so, looked gloomily ahead.

"The knowledge of this ominous future raises numerous questions that are difficult to answer at this time," Dave said in a flat voice.

"Then it will become a complete tragedy. They will place the future of us *earthlings* at their will and determination." Jonathan groped for his pipe, that lay somewhere on the floor.

"Looks like it . . . unless . . ." he said with an almost desperate look in his eyes.

"Hopefully we might prevent that disaster," Dave added.

"It will be a severe mental battle."

Jonathan chuckled. "All our great advanced weapons are useless in the war against this nefarious intelligence from space."

"How does one explain this to politicians, clerics and scholars? They will not believe any of this."

"We should not expect any support or cooperation from that side," Dave growled angrily. For the time being, we're on our own."

A spark appeared in his dark eyes. "No, not really. There is help from the other spiritual world. But that kind of help will not be thrown into our lap just like that.

It's a question of understanding, among other things, the mysterious sounds you experienced while you were in that trance."

With a gravelly tone in his voice Dave Long said: "Indeed. In the trance, when I saw *Nin-khursag*, I was overwhelmed with those mysterious sounds, which I am still trying to interpret in our language. She gave me directions, which I still cannot grasp."

Both men remained lost in thought, staring into space for some time. The tea in the half full cups was cold, but they did not even notice.

"You have to go back into a trance. That's the only way to find out the mystery behind those sounds. You should, when the time comes, concentrate heavily on *Nin-khursag. You must visualize the image you have of her in your mind.*"

"Focus, go back to that moment in the trance." The shaman grabbed the small drum with the eagle feathers. His fingers gently began playing the instrument. Monotone sounds slowly started filling the tepee, like air being pumped into a vacuum.

Dave assumed a relaxed position and slowly closed his eyes. He tried to go back to that moment. He did his best to bring to mind *Nin-khursag's* image. But he did not succeed. He was about to give up, when suddenly his head started to feel very light. Pleasant vibrations began dancing incessantly throughout his body. Then, accompanied by light taps on the Indian drum, which seem to come from afar, he sank painfully slow into a black void. Suddenly he was overcome with an anxious feeling, that gradually faded, as he began to perceive colored flashes of light that became brighter and brighter. The pleasant vibrations again started dancing steadily throughout his body. He felt how he escaped the heavy matter of his physical body. He felt like a green leaf, being moved by the wind and thrown up. The flashes of light changed into a breathtaking display of colors in which red, orange, yellow and finally light green dominated.

The green color began to fade slightly, then unfurled in his mind into another image. It was the beginning of a flight through no man's land, where colorful landscapes full of multi colored flowers slowly passed by.

Then, his mind stopped. He became aware of the presence of something beautiful and sublime . . . Dave Long, at that moment, became overwhelmed with sounds intertwined in an unfathomable heavy symphony . . . The sounds carried a message, which he vainly tried to understand. Suddenly he saw an enormous stone edifice looming in a wreath of the most amazing colors . . . It was a high gate on which many mysterious figures flanked a proud divine being. Like a magnet he was drawn to it. It loomed grand as he swiftly approached it. The large impressive figures engraved in red stone, flashed in his mind.

The evidence was clear!

There is where he had to go to find the key to reversing the danger that had been unleashed on mankind. The symphony ended abruptly . . . unfinished.

He floated downwards, back to his physical body.

Dave opened his eyes, closed them again and opened them once more. The shaman looked at him with undeniable curiosity.

After awakening from the deep trance, Dave tried to convey to the shaman, in appropriate wording, the message contained in the sounds, when suddenly the latter interrupted his hesitant word flow. Surprised he looked at Jonathan, who stared past him.

"I'm sorry to interrupt you so abruptly. Do you hear that too?"

Surprised Dave pricked up his ears.

"These sounds are different from the ones I heard in my trance," he whispered excited. The shaman peered at the tarp that closed the entrance of the tepee. The sound slowly glided closer, it seemed vaguely medieval, like from a classical piece, performed with non-earthly instruments.

The sounds suddenly filled the entire tepee. They were ominous and frightening. "Holy Manitou, this is the symphony of the other side."

"The spirit of *Enlil* lets itself be heard!" Jonathan shouted.

"This will be our first fight against evil!"

No sooner did he utter those words, than they heard how otherworldly instruments changed the dark symphony into a chaos of sickening sounds. It was so malicious in nature, that Dave began doubting his mental faculties. Sounding through this horror that inevitably was coming at the two men, was a bloodcurdling, crazy and unarticulated scream as only a dumb-beaten lunatic in mortal agony could express.

The bone-chilling scream repeated several times, while an incalculable evil atmosphere began to tangibly manifest itself, accompanied by a terrible disgusting stench. Dave Long was overcome with deep suffocating fear. He stood there shuddering, powerless to do anything. Drops of cold sweat sparkled on his forehead.

He began reeking of sweat caused by pure fear, as he turned his head to the shaman who, showing no fear stood in front of him with outstretched arms and clasped hands. On his face with eyes closed, was a grim expression, while he spoke loudly incomprehensible American Indian words. It looked a lot as if Jonathan was involved in a titanic struggle with something sinister. The old Indian sank to his knees, his head proudly raised.

The struggle lasted several minutes, which seemed like hours.

He began singing in a conjuring Native American tongue. Then in a fierce voice, he started speaking aloud, spells in Navajo.

The words visualized in Dave's mind like heavy sledgehammer blows, that pounded incessantly on something . . . *bang . . . bang . . . baaaaaaang!*

Suddenly amidst the weakening chaos of sounds, they heard that frenzied cry again, this time tired and very weak. The sounds moved outside and became weaker and weaker in the distance until nothing could be heard anymore.

Evil had retreated.

There was a deep silence in the tepee.

The shaman got up, visibly exhausted and heaved a deep sigh.

"Wow that was the most severe mental struggle I've experienced, since my initiation many years ago when I was a young man."

"Holy God, what was that? Jesus, and that stench, awful it made my stomach turn completely." Dave took out a handkerchief, sighed, and dabbed his sweaty face. Jonathan filled a glass with water from a jug and took a sip. A worried frown appeared on his face.

We are both witnesses of the struggle between *Enlil* and *Nin-khursag,* which is now clearly manifesting itself on Earth. A battle previously fought on this planet and continued in that other dimension has now been moved once again to our physical world." Dave nodded. "I begin to understand now. *Nin-khursag,* who in ancient times, loved the humans she created and her spouse and brother *Enlil* who couldn't care less about mankind, are at each other's throats again as usual, with mental and spiritual powers unknown to mankind.

"It is his *spirit* that has impelled the feeble race to inhabit new fresh bodies on our planet."

"But you'd think that beings with untold far advanced technological means in their super civilization, would also have high ethical standards and would leave us *earthlings* alone," Dave said angrily.

"Advanced technique, technology and civilization do not always go hand in hand with advanced ethics. In this particular case, it's also about the survival of a species," Jonathan replied, sighing deeply.

The shaman remained calm, but could not avoid the clearly audible emotion in his voice.

"Oh . . . Nin-khursag, Lady of Life, dear mother of us all, please help us and give us strength to turn the tide and win this fight!"

There was a brief moment of silence.

The shaman looked at him sternly.

"Dave you mentioned just now, when I interrupted you, a stone edifice—a gate bearing those mysterious petroglyphs."

"Yes I clearly saw that high gate. I can't be mistaken. I once saw a photograph of it in an archeological magazine."

"I know what you're referring to. There's only one gate on earth that fits your description." He added something unintelligible in Navajo.

"Hold on a moment!" Jonathan quickly stepped outside, leaving behind a bewildered Dave. He shrugged surprised. After a few minutes Jonathan returned inside the tepee, sighing with a book in his hand. "Great writer, this author, his name is Robert Charroux."

Dave sat on the floor again.

"Just a minute," he said as he leafed through the book.

"Hmm . . . Yeah, here it is!"

He handed Dave the book with on the open page a black and white photograph. Dave read the caption aloud: *"Gateway of the Sun (Puerta del Sol) in Tiahuanaco, a more appropriate name would be Venus Gateway. Chapter III."*

Slightly shocked and fascinated, Long stared at the photograph.

"Yes! This is the structure I saw in the trance with one difference though . . . it looked like it had just been built."

"Holy Manitou, in the trance you also traveled beyond our time. Jump to chapter three and read the section about this Gateway of the Sun." Dave flipped through the pages until he reached that chapter.

It says: *"The ruin city of Tiahuanaco is situated at an altitude of 3825 meters."* He skipped a few lines. "I'm going to read now what you've highlighted. *One of the theories states that during the terrible rains of the Great Flood, the subterranean city would have been flooded and silted by loosened mud and soil. This could be an explanation for the fact that this Gateway of the Sun would give access to the void of an absent city. And here's the kicker. The petroglyphs on this Gateway of the Sun probably symbolize interplanetary machines, as evidenced in the documents of Garcilaso de la Vega. The ideogram on the head of a figure is a terrestrial spaceship. It is the image of a jaguar head, signifying power and earthly life. We also see stylized cones, suggesting cabins and residences. If you look further, you'll see the head of a condor, which is interpreted to be a journey through the universe. Many scholars agree that the drawings on the figure represent a motorized spacesuit or—and that is more likely—a repulsion motor. The engine power was undoubtedly used to bring about dissolution or disintegration of the sun in their two polarities, just as they dissolve in the six colors of the spectrum."*

"Thus, an explanation of these petroglyphs is not that difficult if you study them well. The jaguar is a powerful beast, symbolizing the strength of some events in nature and in this image, the ability of a device to move through the universe. The stylized cones are a clear representation of spacecraft we call UFOs in our time, that travel through space and arrive on our planet with on board astronauts, whom the ancients called gods, because of their great knowledge and wisdom."

He paused—a deep frown appeared on his forehead.

"If you read on," Jonathan said "You will come across something interesting about the beautiful *Orejona*, who according to Inca traditions, would have been the mother of the first people on Earth."

"It begins to dawn on me that it is the same female figure as the *Sumerians'* . . . *Nin-khursag, Lady of Life.*"

"We are on the same wavelength," the shaman said hoarsely. You can clearly see the connections and similarities, mangled in writings after all those thousands of years.

Preparations for a long journey took several days to complete. Tiahuanaco was far away. By using the Internet, the two men managed to find the appropriate flights for the trip to Bolivia.

During the flight Dave pulled out a newspaper he had picked up a few hours earlier. He scanned the front page with his eyes. He was a little shocked when he read the headline of the main article: *"Nightmares and suicides increasing"*

Throughout the world numerous people were suffering from horrible nightmares, resulting not only in insanity for some, but also ending in suicide for others. This plague, which assumed increasingly worse forms, had caught the attention of various governments, who requested that the United Nations' Secretary General Ban Ki-Moon, address this matter. This world organization had called an urgent meeting for the following Monday to discuss this bizarre phenomenon dominating the headlines and forcing all other news to the background. Another headline also grabbed his attention. *"Employee of controversial governor found dead."* The editor referred to a detailed article on the next page of the newspaper. Alarmed, he opened that page and continued reading.

The wife of the governor of California was the first person who noticed that her spouse had undergone a personality change, after experiencing the most horrific nightmares.

This woman, Elizabeth Pain, had been interviewed extensively by the journalist. After his discharge from the hospital where the man had been rushed, he appeared to be speaking in a strange unknown tongue, did not recognize his wife and children anymore, and had no understanding of the important documents he had prepared a month ago, which required his autograph. The next morning the man took a series of odd measures that made no sense to anyone. This happened after he received a visit from an unidentified man. They had spent many hours at his office and he was not available for anyone during that time. After the stranger's visit, Governor William Pain again started speaking fluent English with a strange non-American accent.

The paper promised to have more news the next day, after the governor stated to two of his key advisors that he had big new plans for the state of California. Plans which had the two men terribly worried, because of the unusual scope.

The effects would not only severely change the future of this state, but could also have a tectonic effect across the country, if no countermeasures were taken. The two men had left the office, angry and frustrated, after heated discussions with their boss.

Henry Watson, the principal advisor to the governor, who could no longer keep the plans to himself and had gone to the newspaper, was found dead at his home two hours later. The doctor who was summoned declared that death had resulted from "cardiac arrest." The victim's wife found that difficult to accept, since her husband was extremely healthy, exercised daily and never experienced heart problems before.

"Dear God, read what it says in this paper. It has begun . . . !"

"What has begun . . ." Jonathan looked shocked when he almost snatched the newspaper from Dave's hands. With a deep frown he quickly began to read. When he finished reading the articles and photo-illustrated reports, Jonathan said in a hollow voice: "Holy Manitou, we need to hurry. They managed to make the transition successfully."

"Who do you think that unidentified man was who spoke to the governor, I have my suspicions, you know."

The shaman said with a somber tone in his voice: "That guy is obviously one of them, one of those horrors, who managed to take possession of that man's body. And now they are busy making their contacts."

"Dear God . . ." Dave whispered.

"They are clearly aiming at the leaders of the country."

"If they succeed, it may be the end of this civilization."

The White House was flooded with messages via email, Facebook, Instagram, Twitter, telephone, etc., demanding that all psychiatrists, psychologists and those with knowledge of dealing with nightmares be involved to address this problem in a professional manner. Radio and television programs with background information following the news, were devoted to these nightmares and their dramatic consequences for many. Authorities, scientists, government leaders, psychiatrists, psychologists and science-fiction writers joined in, all with the most absurd opinions about this phenomenon.

Clyde Thompson, a prayer healer, who had a powerful and popular radio and television station at his disposal, claimed that these nightmares were God's judgment upon a sinful planet. His services, also available on the Internet and via satellite transmission worldwide, were received by troubled listeners and viewers. Churches were filled with desperate believers. Church parking lots were not able to process the cars of the numerous visitors. Dave and Jonathan watched a reporter on television who had attended a church service, talk about people reeking of fear with big frightened eyes in their pale and taut faces. He also reported about the air conditioning system trying vainly to eliminate the heat and humidity caused by the packed sweaty bodies. The gospel choir sang loudly, but the noise sounded hopelessly impure. Their rock and trust in the Lord began falling to pieces.

Panic among the population increased every day, as more and more people were affected by the most diverse nightmares. The corollary was a slowly crippling economy, since the people involved in the workforce dropped out in droves. The number of suicides grew shockingly high. Hospitals and ambulance services could not handle the situation. They also had to contend with people, who after waking up from those nightmares, had undergone a personality transformation, causing consternation among close relatives and others whom they dealt with on a daily basis. The newspapers were full of horrifying stories on this subject.

The optimism the president had displayed in a television broadcast, when he declared that experts and other scientists were diligently searching for a solution to this serious problem, melted like snow under the sun, when it leaked out that the vice president, after a critically ill period filled with nightmares, had completely gone mad and had to be hospitalized. They began worrying about the president and the other members of his cabinet. The previously always smiling, powerful man in the White House now had a constant worried look. His security detail was reinforced.

As if that would help—this is a deep parapsychological problem, requiring a completely different approach, sneered *The Washington Post* in an editorial.

SIXTEEN

After their arrival at the airport in La Paz, they traveled by train and then by bus to their final destination. The landscape unfolding before their eyes was fascinating—the beginning of an arid desert with its typical scattered thorny plants, colored red by the desert sand. A scorching sun drew gold bands on the clouds, leading into the dark outlines of the mountains in the distance, where the sand and stone-covered Pampa extended and where the sparse vegetation had completely disappeared. Tired and stiff from the long journey, they got off at a bus stop, where they were welcomed by five friendly smiling tour guides with flyers of hotels and places of interest. One of the guides, a friendly young Native American, approached with a big smile and greeted them in English. They immediately decided to go with him. Chatting sociably, he took them to a small hotel, where they took up residence. The hotel was fresh and clean, with a friendly staff consisting of a heavyset girl and two burly short men. It was 04:30 p.m. local time. They decided to take a shower, have a bite to eat, rest for a while and then climb up the mountain where the ruins were.

They stood a few yards from the awesome *Gateway of the Sun* high in the mountains in the ruin city of *Tiahuanaco*. An hour ago the sun had set in a power display of golden fire. The hands on Dave's watch stood at 07:55 p.m. High above, stars twinkled in an inky sky. A crescent moon in the east faintly lit the surroundings with a tiny yellow glow. It had been a strenuous climb up.

Tiahuanaco always makes a deep impression on its visitors. Astounding, amazing and incredible, was the ruin city on the mountain, as if it had been flung down loftily by a natural disaster with titanic power.

Massive heavy stone blocks of fifty and sixty tons each had been placed on hundred-ton sandstone boulders. Rocks, flat and smooth with sharp grooves, stood beside gigantic stone blocks connected by copper clamps, turned dark green by the test of time. They also saw rust brown stone blocks with sixteen-feet-long portals and eroded tiles carved from a single block of stone.

As they laboriously continued their trek, plagued by the cold and thin air, they also noted a design of meandering stones, presumably water troughs of six-and-one-half feet wide and about sixteen feet long, built in the ground. During their ascent they also saw the ruins of a pyramid-shaped temple.

"According to the experts, the latest research shows that this temple which controlled the city in ancient times, was not destroyed by a disaster, but had not been completed by the unknown builders," Jonathan said with a slight smile.

With a straight face, Dave Long looked at the sharply defined contours of the millennia-old edifice. The shaman could barely conceal a slight emotion, when he let his eyes go over the enormous legacy of the ancient race that lived here in former times.

"So, at this *Gateway of the Sun* we will find the final part of that mysterious symphony," he whispered excitedly. A long silence ensued. Lonely sounds of nocturnal animals filled the darkness surrounding the two men.

"Sounds containing a message only you can understand," Jonathan said in a hoarse voice. They walked confidently forward and arrived after a few minutes at the courtyard. There stood, imposing, the purpose of their journey, the *Gateway of the Sun*.

SEVENTEEN

Silent and lost in thought, they stood in front of the mighty *Gateway of the Sun* and its secrets. Was the key to the final part of the unfinished symphony hidden in these petroglyphs?

Their guide, Juan Castro, was a friendly young Indian with a stocky appearance and tough calf muscles, which he owed to frequently climbing to this altitude.

The tourist season was over. Hence, they had the place all to themselves.

They were so curious and excited to see the *Gateway of the Sun* that after merely an hour's rest and setting up their tents, they immediately headed to the purpose of their arrival. They planned to spend the night there. Dave had a photo and video camera with him. He hoped to take beautiful shots.

Jonathan directed the light of his flashlight to the top. "Hey, wait a minute, go down for a second, to the right, a little more," Dave shouted suddenly excited. "Look! I saw this huge in the trance!" He felt his heart pounding, as he stared at it with wide eyes. They saw a mysterious figure engraved in the stone. A condor, in his hand an object with two tubes, was gazing at the sky. Numerous incomprehensible figures and objects were displayed on the condor. Long took several pictures. After staring at the *Gateway of the Sun* for a full hour, in particular the figure he had most clearly seen in the trance, Dave and Jonathan left the site in an exciting mood.

The next morning Dave woke up early. He had only slept a few hours.

In spirit, he still saw the mighty structure looming in front of him, wreathed in glowing multi colored flames. It was an unforgettable magical moment when he saw the magnificent petroglyphs up-close in the trance. Especially the image of the condor with that mysterious figure spoke strongly to his imagination.

The sun was attempting to rise in the east in a chaos of orange red fire. Musing he watched. The golden sunbeams played faintly on the ruins of the mysterious city of *Tiahuanaco*. Abrubtly he heard the shaman's voice behind him. "The Incas and Mayas say that this city was founded by the gods" . . . Contemplating, he added . . . "Aliens or maybe an unknown ancient race . . ."

Jonathan was standing beside him now and looked totally captivated at the ruins of the temple city.

"Yes, my friend, those strangers have done impressive work here, with machinery and equipment completely unknown to us."

"I consulted some literature before we set out on this journey. The *Gateway of the Sun*, one of the archeological wonders of this continent, was carved from a block of stone." They remained standing for a while, watching and philosophizing about the glorious past. The entire surroundings breathed an atmosphere of mystery and tranquility. Dave felt a void in this place that testified to a virtually unknown and obscure past, about which many scholars bombarded the world with all kinds of clear and even more so vague theories.

"We're going to freshen up quickly, have something to eat then return to the *Gateway of the Sun*." It was bitterly cold at this altitude. The breath of the two men, as they stood there, was as mist in the air.

For the second time, since their arrival at this site, they were guided to the *Gateway of the Sun* by Juan. It was a cold but beautiful morning. A clear blue sky with low-hanging white clouds in the west typified this Thursday morning.

Occasionally a strange feeling came over Dave, that this long journey was a hopeless task and would come to naught, but when he thought back to the trance and the other incidents he had experienced in a short time, that rotten feeling somewhat disappeared into the background.

In the past few days he had also developed a deep respect for Jonathan, an initiated shaman with an intellectual background—that was something special. Jonathan was a very distinctive man with a broad general spiritual view of things in everyday life. This old Indian was as cool as a cucumber and was not easily distracted by unusual incidents. And that was currently very important, especially in view of the forthcoming shocking events.

Events, if not reversed for the better, would have a devastating effect on humanity.

Leisurely they walked on a wide improvised red brick path towards *the Gateway of the Sun*. Juan the guide, in excellent English, gave both men an explanation of the environment and his experiences with the numerous tourists visiting this site. Some of them come here with religious and other strange intentions. "The ruins of the temple city speak strongly to the imagination of many visitors, who come here from all regions of the world," Jonathan commented casually. After fifteen minutes, they stood once again in front of the impressive *Gateway of the Sun*. "The height is about nine and the width thirteen feet. I have read that the weight of this mighty edifice is estimated at ten tons," Dave said with narrowed eyes.

Magnificent in the rays of the rising sun, the *Gateway of the Sun* displayed an impressive figure, representing a flying deity flanked by three rows of creatures with strange helmets and headdresses in squares.

In the neatly restored courtyard they also saw a collection of stone heads. Some of these faces had strong features. Others were portrayed with soft features. There were faces with full and others with thin lips, with long and curved noses, with dainty and plump ears.

"Take a good look," Jonathan said. "What we have here is a collection of heads, of the most diverse representatives of human races. Look at those faces."

"Wow, Dave nodded surprised. "Impressive!" They walked closer to the *Gateway of the Sun*. "Look at those figures flanking the flying deity. Do you see the one at the far left bottom? Last night we saw it too. I saw it in the trance . . . I was standing right atop in that trance state . . .""

Dave snapped the camera several times.

"These will be great shots," he exclaimed enthusiastically.

"Hey, how can that be?" He heard his friend say suddenly.

"What's that old woman doing here?"

"Which old woman, where is she?"

"Look to your left . . ." whispered the shaman. Dave looked discreetly to the left. An old Indian woman dressed in typical garb indicative of Indian women of these regions: long skirt, thick sweater, scarf and felt hat on her head. "What is so special about that woman?" Long asked slightly amused.

"First of all, how did this woman get at this altitude? And secondly, I swear on all sacred Indian graves, it is the same woman who sold me that statue on the flea market." Now it was Dave's turn to be dumbfounded.

"We need to bring clarity to this mystery or clue of something very important.

The shaman hissed through his clenched teeth. "Come with me." He walked towards the woman who was standing between two of the stone heads. The light of the early morning sun shone in their eyes and with the greatest difficulty they tried to follow the woman who slowly disappeared between the two heads. Quickly they walked over to the spot where they had seen her standing. "Where's she gone?" Dave growled. The shaman said nothing and kept walking. Suddenly he stopped, looking surprised. "I have a strong suspicion of what's happening here."

"What do you mean?" Dave asked while stroking his stubble beard. "Look at the behavior of our guide," Jonathan whispered. Dave glanced back. Juan stood at a distance staring wide-eyed at the spot where they both were. One could clearly see that he looked agitated and afraid. Dave walked towards him. "Why are you so scared?" Juan shook his head and pointed forward. "That old woman . . ." He paused suddenly, terrified.

"What about that woman?" Jonathan came closer. "A week ago she was here too. I was accompanying two British people. Those white men did not notice her while taking pictures over there." Juan pointed in the direction of the *Temple of the Sun.*

"This woman appears out of nowhere and disappears from your sight before you notice . . . weird, huh?" Dave, who said so, was shivering.

"So you didn't see her earlier?" Juan shook his head anxiously. "No sir."

Jonathan looked at Long. "I said just now that I have a suspicion of who she is, or who she possibly could be." Jonathan scratched his chin. "I'm sorry, but I have to keep that to myself for now—I must think."

Long shrugged.

"Another mystery. First you met her at that market and now here." Again the shaman looked in the direction where the woman had appeared. He clenched his jaws and the fires in the back of his eyes began burning brightly. "This appearance, this old woman wants to tell us something and undoubtedly she also plays an important role in the overall events that have occurred around us lately."

Dave's thoughts shifted again to the *Gateway of the Sun.* "What I don't understand is that figure on the *Gateway of the Sun* from my trance, that I have now seen in the flesh, but which still doesn't mean a thing to me," he said wearily shaking his head displeased.

"You will," the shaman said mysteriously.

EIGHTEEN

Slightly discouraged and disappointed, Dave and Jonathan made the descent with Juan. He had not gotten much wiser, after seeing live what he had seen in the trance days ago. The guide took them to their hotel, where they gave him a big tip on top of his pay. "Contact us again tomorrow morning," said Jonathan when they parted.

That night, when Jonathan projected the images from the camera on a bed sheet secured on the wall, they saw something strange. "Hey stop!" sounded Jonathan's voice. "Do you see that?" On the image appearing before their eyes, they saw grooved in the stone, a distorted cruel face staring at them, angry and ferocious.

There was fear in the eyes of both men.

"*Enlil*," moaned the shaman.

"But, how's that possible?" At the *Gateway of the Sun* we saw something else in that square."

Pensively the two men stared at the still image on the sheet.

"Enlarge that image!" Dave immediately followed the other man's order. They looked at it, again. The image was three times larger than before. "Those lines . . . those lines . . ." Jonathan muttered.

"What the hell is wrong with those lines?" Dave asked.

"Look to the left . . . there is something, beside the face."

"I see nothing special," Dave growled uncertain.

"Look better man!"

"Only now did he see something.

He took a closer look. His jaw dropped in shock.

"This is f . . . ing impossible . . . Good heavens how's that . . . ?"

Ann Blackson was having a difficult time. After the departure of her two new friends, she was constantly troubled by a severe uneasy feeling. She had already received three email messages from Dave, telling her that everything was going well and yet, that feeling had increased in intensity this evening. She tried focusing on other things, but to no avail. Deep inside, an anxious feeling was soaring, concerning both men in Bolivia. She knew with all certainty that if there would not be a timely intervention, things would turn out wrong for Jonathan and Dave.

Nin-khursag, the beautiful woman, had once again appeared in her dream. What had stuck in her mind was the sad and pitying look in those almond-shaped eyes staring at her. It seemed that night, as if this apparition stood beside her bed. It lasted only briefly. She woke up and was immediately overcome with severe headache. That was four nights ago. Since that moment, she had not been able to shake that uneasy feeling, it only got worse. At work, she had difficulty concentrating, and that awful headache kept bothering her. Her boss noticed the always smiling Ann had turned into a skittish woman, occasionally looking over her shoulder. He could no longer stand it and advised her to see a doctor, which she did that same day. The doctor, familiar with the prevailing phenomenon, spoke to her for some time—it so happened, that he had listened to her interview on a local radio station. She evaded certain questions he asked and kept the conversation light on what she had told the interviewer during the program. He gave her a week of sick leave and also prescribed something for the headache.

–Why should I tell him everything, he would not believe me anyway and, besides, he would not be able to make sense of it—she thought as she got in her car and drove off with a cold smile on her lips.

When she made the turn into the street leading to her house, she saw someone standing in front of the fence. She frowned. "Who could that be," she wondered surprised. She stopped and got out of her car with a suspicious feeling. The stranger, a tall white male, neatly dressed in a two-piece brown suit, looked at her and smiled. He greeted politely. "I have been waiting for you for five minutes."

"For me? But how do you know that I would be here at this time?" A thousand confusing thoughts were going through her head.

"Don't be afraid," the man said with a smile. He held out his hand. "Eric is the name, Eric Bancroft."

"Ann Blackson," she answered hesitantly. "I'm here with a special purpose," the stranger said. Ann did not like the tone in his voice. It sounded anything but friendly. "Oh yeah!" She became aware of something malicious radiating from the man. She wanted to storm into her house, but it seemed as though her legs refused service. "Why are you so nervous lady?" He asked with a kinder voice. She looked at him suspiciously. She suddenly broke out in a sweat. She exhaled deeply. "Why are you sighing lady?" Not long ago, you were able to escape . . . those intruders. Your strength is deep within you, but . . ."

"What do you mean, and . . . Who are you?

"Take it easy, you will find out in a moment."

He looked at her with his piercing grey eyes. Only now, did she notice the grey color.

"I will not keep you much longer. Remember, what I'm about to say."

She stared at him, still impartial.

"Your two friends will never come back. We will take care of them there in that foreign country." He gave her a dark, malevolent look.

She was terrified, felt her heart pounding wildly and her face turned ashen. "Who the hell are you and what do you mean?"

"I thought I was clear," the stranger said mockingly.

"With an undertone of menace in his voice, he added: "Do you love Jimmy?"

"How do you know my little boy's name? She asked surprised.

"We know everything about you and your two friends. Nothing will stop us on our chosen path."

"Stay the fuck away from my child!" she began screaming hysterically, while tears of rage flashed in her eyes.

"Keep quiet! Therefore, do not contact those two, if you cherish your life." He paused briefly. *"And that of your child . . . !"*

"I greet you."

Ann watched his vanishing back, as he retreated.

Her eyes betrayed the great dread washing over her at that moment.

NINETEEN

It was a complete mystery. Once again, Dave studied the enlarged image projected on the bed sheet. That was unmistakably his face. There was no doubt about it. Flabbergasted he looked at Jonathan, who said nothing; there was a deep frown on his forehead. After a long while, the shaman began to talk. "Now I'm really stuck." A deep sigh escaped his lips. With a grim look on his face he said: "This does not only require concentrated thinking, but also meditation, better yet, you have to go into a trance."

"And when should that happen?"

"Right now," the shaman said firmly. His voice was hoarse.

"But I still have to fast . . ."

"We don't have time for that. In a short time, the most unbelievable and strangest things will come at us, screaming for answers. *Enlil* is maliciously engaged. We must stop him!"

Dave felt himself sinking again into a non-physical world. Flaming colors came at him. In a flash he saw in this trance state, the silhouette of a dark figure in front of him—threatening and massive. He fearlessly flew past. The disembodied state in which he found himself, made it seem as though he was being propelled by a fierce storm. He felt like a butterfly that could be blown away in another direction at any time. Unperturbed, bundling his spiritual powers against an unknown enemy force wanting to overwhelm him, he flew further and further. The fact that he was going somewhere was obvious, but his final destination was completely unknown to him. Beneath him an inky sea, whose wild waves posed an unfathomable threat, shot past.

Abruptly the storm died down. He arrived in a world unfamiliar to him. Below him, a beautiful beach with the sea very far. It was a wide beach with at a distance some swaying coconut and other palm trees. Behind the palm trees he saw some dunes covered with green grass and others with moss and flowers. He landed in standing position on the white sand that crackled under his bare feet.

Surprised he looked at his feet and then at his light bronzed skin. His hair fluttered in the wind. To his astonishment, he only wore some sort of loincloth, made of rough brown material. In disbelief he looked at his muscular arms and legs.—Holy shit, who am I in this physical body? And what am I doing here in this unfamiliar environment? Overhead was a cloudless clear blue sky. He looked around indecisively for a while then opted to walk over to the nearest coconut tree plantings. He did this also, because the bright sun began scourging his skin with its hot rays.

–Stay calm dude . . . keep walking, Dave encouraged himself, as he continued walking. He had come within a few yards of the coconut trees and dunes, when he noticed the strange bird. The beast let out a strange scratchy sound, flew in circles above his head flapping his wings, then disappeared behind the hills. Dave felt uneasy. This was a creature of a bygone era. That spiny head and the leathery wings spoke volumes.

He froze.

–Dear God, did I land in a prehistoric era?

He wondered curiously what he would find behind the hills. As he approached, Dave saw that the grass and moss was interrupted by brown and red blocks of stone which sparkled under the sun. He also saw scattered thorny bushes growing amid rampant weeds. Numerous colorful butterflies flew up and down above the greenery. An oddly shaped brown lizard scurried away when he came nearer. Now he had to fight his way through a large number of high-growing shrubs with white and red flowers. Two hummingbirds, with their rapidly moving wings, flew from flower to flower.

As he continued walking, Dave heard a rustling sound behind him. He turned his head. Out of the corner of his eye, he saw something move. He looked closer. Something like a big dog was watching him wagging his tail. It looked like a dog, but when Dave took a closer look, he noticed that this beast still showed a big difference with similar animals he knew. The beast's head was rather reminiscent of that of a bear species. It had a dark grey coat and looked at him curiously. Suddenly he heard sounds from afar. Those sounds were unmistakably human. The beast perked up his ears, made a growling sound and disappeared into the bushes.

Dave heard someone coming. He wanted to hide, but it was too late—a female came from the high growing thicket. Surprised, she froze with a startled look on her face. Instantly a smile appeared around her pearly white teeth. She was a lovely oriental type; Indian-like. She walked briskly towards him, the wagging dog in tow. She bowed her head and greeted him by holding her palms together, as he remembered the Hindus do. She said something, probably a name. It sounded like "*Tong . . . Tonghh.*"

The girl was scantily clad, like him, in a loin cloth and a slight top, barely covering her beautiful breasts. She also wore a string with colorful beads around her neck. Only now, did he notice that she was wearing brown leather sandals. She began talking to him. He was shocked, as this was not the normal speech he was accustomed to. Sounds were coming from her mouth. Sounds that began to caress his ears like a song.

–Good heavens, in what kind of world did I arrive? He thought nervously. Suddenly it occurred to him that in the trance *Nin-khursag* also spoke to him in sounds. It seemed so long ago.

The girl pointed forward. He understood her immediately. They had to go in that direction. She said nothing further and looked at him with a constant smile around her open mouth. Occasionally the growling dog ran behind them.

This girl looked so familiar to him. He racked his brain tired. She was no stranger to him.

—I know her very well, too well. How the hell was that possible? My God, from where? After a while he gave up.

They walked about fifteen more minutes and arrived at a well-kept area of coconut trees, behind which Dave saw some large thatched huts. The surroundings were remarkably quiet. Not a human in sight. He followed the girl, who nonchalantly walked past the huts.

Another fifteen minutes or so later, he heard faint sounds carried by the wind from afar and the noise of hammer blows. There were people at work over there. They arrived at a plateau looking out onto ruins of enormous structures. It seemed as if a giant had smashed everything. All around him were pieces of stone churned up walls, buildings and other unidentifiable remains of once enormous structures. The ground was ruptured in many places.

They continued by following a narrow path. As they kept walking, he came to a shocking discovery.

This place looked familiar!

Squinting from the sunlight, he looked around. *"Good Heavens, this is indeed Tiahuanaco!"* He whispered excited.

They found themselves on a red stone plain, where in the distance men were at work. He continued to follow the girl. They took another path lined with rough yellow green bushes. Before them, a colossal stone artifact, several yards high, came into view. They arrived, once again, on flat terrain with here and there some vegetation. He was horrified by what he saw directly in front of him. Hoarsely he said: *"Dear God . . . The Gateway of the Sun!"* Quickly he and the girl ran closer. Curiously, he stared at the *Gateway of the Sun*. It was an overwhelming sight. The *Gateway of the Sun* sparkled in the sun with a nice soft shine. It seemed as if the structure had been built yesterday. It looked new.

Dave Long was overcome with a deep sense of respect and reverence. Before he realized what he was doing, he sank to his knees.

Through his eyelashes he saw the girl doing the same. A few soft sounds escaped her mouth and she stood up again. He followed her example and wondered,—what now? As if she had read his mind, she took him by the hand and walked towards the men, who had stopped working. Happy and surprised they looked at Long. Surprised he looked at the men.

–These men know me.

Who the hell am I in this body . . . ?

TWENTY

The shaman continued watching the motionless body of Dave, who was sunk in deep trance. He wondered where the white man now tarried. Hopefully, he came back with clear answers and directions on how to proceed. That incident an hour ago was also a complete mystery to him—when we looked at the figure contained in the square, only the condor was visible with the incomprehensible characters. On the image projected on the bed sheet, however, they saw something completely different. Besides that ferocious face, under which the clear lines, was the image of Dave Long. How was that possible? Quietly, Jonathan went to lie on his bed, emptied his mind and concentrated on recent events. It seemed like a large jigsaw puzzle, the matching pieces of which were mixed up. He closed his eyes and attempted to connect everything. In semi-trance, he visualized the moments in his mind and all of a sudden, he saw crystal clear the true picture that showed him everything. It was no more than a brief flash.

He opened his eyes, grabbed a pen and paper and jotted down a few words that turned into sentences.

"Holy Manitou, that's it!" He whispered hoarsely.

And now I have to wait patiently for Long's return.

One of the men came over to Dave. He had deep grooves in his face and was, unlike the others, dressed in a long robe of a brown red fabric. He had a brown hood over his head, whence peaks of grey hair were visible. The man looked sharply at him with his coal-black eyes. Dave was overcome with an inexplicable emotion. The man looked very familiar. He could swear he had met him only yesterday.

The older man beckoned him aside. They went to the side, a few feet removed from the other workers, and stood across from each other to talk. With a serious face, the man carefully took, from his wide coat pocket, a package wrapped in a black cloth. The package was a foot long and as wide. With a slight bow, he handed it to Dave. Curious as to what it could be, Dave opened the cloth. What he saw made him tremble on his feet with excitement. It was the small image of a figure, crafted in dried brown clay, encased in a square, that he had seen projected on the bed sheet in the hotel room. The ferocious face, with the proud and cruel features immediately noticeable, was more than familiar to him.

"*Enlil*," he whispered . . .

The expression on the old man's face changed. He looked at Long with fear in his eyes. The corners of his mouth were tight. He nodded twice.

"*Good heavens how do I make myself understood?*

I have a lot of unanswered questions."

Dave's eyes moved to the bottom of the image. He had expected it, yet the confrontation with his own image stunned him in such a way, that he began shaking and broke out in a cold sweat. When he raised his head again, the eyes of the old man looked straight into his.

"*I Nolok, saw it right away. It is you! You came back and now you have to complete your task.*"

The man spoke to him in sounds, sounds that were translated into words and sentences, which to his shock and amazement, he could hear in his head!

The man uttered a series of sounds that came to him by way of telepathy. It was amazing how those sounds flooded him, and the message came to him translated into words he completely understood.

The shaman was worried. Two days had passed and still no sign of Dave Long's return from the deep trance. Motionless, he still sat in that state. His eyelids blinked occasionally in that taut blank face, void of any emotion.

It was a matter of waiting patiently until Long's *psyche* had returned into the physical body, in front of him. From past wise lessons and experience, Jonathan knew that you always have to wait quietly until the one in the trance gives you a sign of his return. Intervening prematurely, could have dramatic consequences for that person.

A slight smile appeared on his thin lips.—Surely, Dave must be on a very long journey to other worlds, where he, hopefully, would find the right answers and clues to all the questions they had to contend with here in this ancient place.

"If fate has destined a particular person to be at a particular place, then let fate lead that person by his own needs to that place. That is why you are here right now." These words rose from the depths of his heart, while the wise old man addressed him in sounds.

They were now seated on a block of red stone with a polished surface. The other men continued working steadily. The girl, who had brought him here, had retreated with a smile. He understood that she would return shortly with food and drink.

The wise *Nolok* told him about the gods who, a long time ago, had descended from the blue sky and had founded the city, which had now fallen into ruins of enormous stones and incomprehensible objects.

The elders of the people had processed and left behind in many ancient writings and drawings, the purpose of the utensils, weapons and all sorts of *celestial chariots*, with which they were able to move in space and time.

They had, after the usual solemn ceremonies, begun construction of the gate of the sun. After the construction, *The Gateway of the Sun* stood as a confirmation of these great events that took place in the past.

A monumental stone witness of a glorious past, when the gods dwelt amongst men.

During the construction itself, numerous inexplicable events occurred in which *he* had played an important role, with dramatic consequences that would be felt in times to come. The wise men among the people had destined him to travel to a distant future, in order to change things for the better.

He was listening attentively to the wise man, who gave him a clue how to find the key to undo *Enlil's* pursuit of absolute domination over the people on Earth in a distant future.

The wise man, in visions and reflections of the future, had seen a future that looked grim for the people of that time.

"The key is hidden in a sign on this Sun Gate, that you have to find yourself. I can only give you a clue.

Between Enlil's concealed face and yours, a startling secret is hidden, that will have shocking consequences for Enlil's appalling objectives and his sinister ideas.

With this key in hand, you will perform a relentless battle.

Your path will be full of thorns, but if you persist, you will be successful. However, I foresee severe consequences for those who assist you.

But that too can be turned for the better, unless you should fail . . . "

The wise man got up and said in tones that sounded like the murmur of river water against a pebble-littered shore: *"Come, I will show you things hidden from the eyes of man . . . "*

TWENTY-ONE

From the chronicles of Inrith (part I)

After an interesting period of learning and penetrating into a tough subject of knowing the invisibility of things to the normal eye, it was high time that I, *Inrith,* returned to the place of my birth.

Following a lonely trek through the forest lasting three sunsets, I reached flat land, where I heard the sound of the sea. The sound made me walk faster. I had to walk all the way to the end of the beach, in order to reach my hometown.

My thoughts went back to the time when I started my journey, to the place where the men of wisdom lived.

Countless sunsets had passed since she accompanied me to that beach. It seemed like only yesterday. What would she look like now? I often wondered. The always smiling *Mirath* with whom I had grown up and who always stood by me, whenever I asked for her help with some intense issues, to which she always knew the answer. It was my firm conviction that she would grow up to be a wise woman. That was also the view of the other villagers.

With a mysterious smile on his lips, *Nolok* had always watched when the two of us were engaged in profound conversations on diverse topics, pertaining to life in the village. I also thought of the old man who, undoubtedly, was longing to see me too.

As I continued walking, the roaring sound of the sea became stronger. After a while I saw the white beach in the distance sparkling under the bright sunlight—the beach where *Mirath* and I often used to hang out, cooling off in the blue water of the sea.

Once again my thoughts returned to the day when I began my journey, to the place where I would be initiated and receive the teachings.

She walked with me to the beach. With a greeting and a smile on her face, she said goodbye to me. I looked back one more time, before she eventually turned the corner, where the beach made a curve. The small figure in the distance waved at me once more. I waved one last time. With a melancholy feeling I continued walking. It would be a long journey to my destination—a small village, far away from my hometown. The village of the wise men, where as a youth, I would receive the wise lessons. Lessons, which would provide me with the necessary insights into the intricate metaphysics; insights only a chosen few would acquire. I was born with gifts which were immediately noticed when, as a little boy with great show of strength, I rescued my native village from the clutches of a devastating storm.

In the annals of the village, the crippled *Ziat* worded this as follows:

Small and brave with his eyes closed, Inrith walked into the storm with his hands stretched out. The wind, behaving like a capricious child, blew violently for a few seconds and then died down. Stunned, the elders among the villagers had followed the actions of the little boy and after this event they took the child under their personal protection. They taught him all the knowledge they possessed. For the continuation, he had to, when he became a young man, leave for the village of the wise winds where the great Dida dwelt and where nobody came, unless they were in great distress and in need of urgent help.

I walked on the grass that before these hot times had been green and tender, but now looked yellowish brown and was slippery to the touch under my calloused feet. I now reached flat land where nothing grew; on the horizon was the green of a far reaching forest I had to traverse, in order to reach my destination after several days. Physically exhausted, I reached the village where my arrival had been expected.

The wise ones had big plans for me, and after a short resting period of one day, they immediately began with my training and education, which would take several years.

When I finally reached the beach, the joy in my heart enkindled at the ensuing reunion with my loved ones. Suddenly something weird happened to me. I got a strange feeling as if I was being sucked into the air. I was overcome with dreadful dizziness, accompanied by vibrations that ran all over my body. I closed my eyes, aghast by what was happening to me. The vibrations became faster and stronger. It suddenly became dark around me. It was a special kind of darkness that I could see with my eyes closed. I got a strong feeling as though I had stepped out of my body and was sucked into the air by an unknown force coming from elsewhere. The darkness cleared at a certain point, as if the sun broke through and tore the dark asunder.

A bright beam of light pierced the darkness. I began to spin like the spinning top I often played with as a little boy. I went faster and faster until I was no longer aware of the things around me.

Suddenly, with a slight jolt, my consciousness returned. I found myself floating back down in great confusion, as I became aware of other memories that began seeping into my mind.

In standing position I landed on the white sand again. My bare feet touched the sand. *The hot sand under my feet caused an unprecedented strange sensation.*

TWENTY-TWO

The old man walked ahead of Dave Long. His robe scoured the ground; the hem was slightly red from the dust that lay on the ground. Silently they continued. Curious and worried, Dave wondered where they were headed.

They now climbed a path, worn by numerous footsteps. The path went progressively more upwards against the reddish brown colored hills. On either side, unruly thorny bushes colored by the reddish dust, were firmly rooted between the pebbles scattered everywhere. After a while, they came to a gentle curve ending in a massive overhanging block of stone, which hung like a bridge over the path. A few feet further they arrived at a dark cave, carved into the hard rock. The old man made a gesture with his hand, inviting him to go inside. *Nolok* reached for a torch on a rack on the wall, where more of such objects had been placed. He tried igniting it with a flint. The sparks flew everywhere. After three attempts, the torch flared up and illuminated the inside of the cave. The flames were lost in the darkness stretching before him. "We're not there yet," it sounded in his head, when the wise one looked at him.

The flame lit a path and played games in whimsical patterns on the wall and ceiling while they walked. Several minutes later, he heard the gurgling sound of falling water before him. They were approaching a huge space, where an underground waterfall plunged into a creek that disappeared into a dark cavern. Dave looked around in amazement and wondered about the perimeter of the space with high above a dark ceiling of stalactite. He also saw stalactite on the walls of the spot where they were standing. He saw all sorts of unfamiliar objects. On the ground was the hide of a huge bear. The head of the dead beast was still attached to the skin. The wide-open mouth displayed some enormous white teeth, reflecting in the light of the torch. The old man took a seat and invited him to do the same. Quietly, Dave sat down and inhaled the scent of the falling water mixed with the light aromatic scent of the cave. The silence, apart from the falling water, was rather oppressive, he thought. He shook his head to get rid of that feeling. Confident and calm, he then looked at the wise man who closed his eyes for a few seconds and opened them again.

The wise man closed his eyes again, opened them once more, and took an oddly shaped object on a bench next to the bearskin. It was kind of a dark cube made from glossy bluish material Dave could not identify.

–You have to gently open up your mind to the things you are about to see and hear, he heard in his head. *Nolok* opened the cube by chucking it lightly with his right hand. Something popped out. It looked like a piece of dried animal skin.

What is stated on this sheet is what has brought you here—it sounded again in his head. The silence, apart from the falling water, still gave him that uncomfortable feeling. The old man handed him the sheet. He took it in his hands, it felt soft.

–Look at it and allow what it says to penetrate your depths—it sounded again.

As he studied the characters on the sheet, he was suddenly struck hard by a mental probe. He was terribly shocked, recovered from the shock and trembling all over, he looked again at the mysterious characters, which suddenly began dancing before his eyes at otherworldly music, produced on unknown instruments by masters of orchestral art.

This unequaled masterpiece sounded very familiar . . .

–*Good heavens! This is Nin-khursag's symphony!*

The sound reflected from the walls, from the ceiling and from the ground. It engulfed him, hitting him in all the cells of his body. He was submerged as by water from a heavy rain storm, in which he nearly drowned, yet was able to keep standing with the utmost difficulty. The sound continued without a hitch until the end. In his mind, the symphony described all that was good and bad, culminating in an overwhelming crescendo of triumph . . .

Dazed and not yet fully in command of his senses, he sat there with the sheet in his hands.

Then the truth came to him.

On this sheet in his hands was the completed symphony.

With this data, this sheet on which the composition was written in those flamboyant characters, he could destroy the cruel phenomenon that *Enlil* had brought about on Earth, in a time far in the future.

Finally he was awake again. Still a little shaky, he looked at the precious piece of animal skin in his hands. His gaze turned to the things around him and then to the old man, who sat before him with his eyes closed.

Laboriously he climbed the deeply worn-out path that had taken him to the cave below.

The old man walked in front of him. When they came out of the cave, he noticed that the sun had passed its zenith and was moving westwards. In the distance somber grey storm clouds were gathering. The sun was about to disappear and a violent thunderstorm would erupt. Already a few drops were falling down. In his loincloth, he had stored the sheet with the symphony—one end sticking out by his belly. It was a beautiful day, Dave thought, even though it would start raining soon. Nothing could spoil this day anymore, now that he had the complete composition in his possession.

TWENTY-THREE

As they approached the large huts, the pretty girl came outside. She walked down the path to meet them. She was beautiful and enchanting, he thought.

She looked radiant with a hopeful look in her eyes.

He understood her immediately and nodded. Her smile widened. Her smile was so familiar to him. He shook his head, severely confused. For a moment she looked at him in surprise. He just felt it, but said nothing. She would not understand him anyway.

They sat down in one of the huts and ate game and bread. He enjoyed it very much. Water from a brown jug was cool and refreshing.

The time had come to say goodbye to these folks, who looked so familiar to him.

The old man held both his hands firmly. "The best to you my son—go and accomplish your task. It will be alright." With the girl by his side, he walked back to the beach. The dog appeared from behind a hut, kindly wagging his tail.

The rustle of the sea breeze could be heard more clearly, the farther they walked. Through the vegetation, he could see in the distance the glistening beach in the sun. During the trip to the beach, she occasionally had said something in beautiful sounds. In his head it sounded like words: *Our roads are parting, you have a long journey ahead. I wish you all the best.*

After a few minutes they reached the beach. In spite of the calluses on the soles of his feet, he felt again the tickling of the hot sand. She held both his hands. *"Good luck my dear brother."* The sounds rolled from her mouth in a passionate rhythm, as her eyes became moist. He was overcome with emotions, gave her a long hug and walked away. He turned around one more time and saw her standing there; a divine figure bidding him a final farewell, then disappeared into the bushes followed by her dog.

Steadily he continued, wondering—what now?

How would he return to his old environment, and in his other physical body? How long had he been away?

The shaman must be terribly worried. His anxiety was increasing and took on proportions that gave him a slight headache. He wanted to sit down, not that he felt tired, but to think.

He shook his head and kept going.

The weakened sun began to set in a chaos of orange colors. The horizon, where the sea touched the sky in the east, was also colored orange. On his forehead was a deep wrinkle, as he worriedly continued walking.

His contemplation was interrupted when, to his great joy, he began feeling the familiar vibrations go through his body. "Dear God, I'm going back." He wanted to cry out of pure joy. He also became aware of a hissing sound in his head. The vibrations intensified and increased in frequency. It was as if he began to float a few inches above the sand. Suddenly he felt a strong suction from outside, pulling his *psyche* from that body.

He sensed becoming disengaged from the earth.

In a flash he saw the body he had just left, was now inhabited by the other one, who continued walking on the beach.—*Oh God, I lost the sheet with the completed symphony*—was the last thought flashing through his head, before he sank into oblivion.

Dave felt that he was somewhere, lying on his back. He also had the sensation that his face was being dabbed with lukewarm water. He listened attentively—the sound of a voice. Someone was singing. Unfamiliar words contained in a monotone song. The melody was being repeated over and over again. The desire to open his eyes seemed to have abandoned him. He tried again with great effort of will.

Finally he succeeded.

Dazed, he looked into a sympathetic brown face that smiled at him kindly and concerned. Slowly he figured out where he was. "Oh God, Jonathan, I'm back, ouch, my head!"

Dave sat up. His head throbbed as if someone had worked it with a hammer. "Welcome back, relax a little. Pull yourself together. Would you like something to drink?"

"Yes, water, preferably ice cold!"

Jonathan walked over to the small refrigerator and took out a bottle of water.

Long drank in big gulps. "My throat felt terribly dry. That water is delicious." Still a little groggy, he sighed deeply. "Here's a tablet for your headache," the shaman said. Dave ground the tablet between his molars and took another sip of water.

Suddenly the memories of the things he had experienced came rising to the surface with a jolt. Overcome with panic, he groaned frustrated: "Jonathan, it's awful, I lost the symphony. I . . . I . . . He walked away with it!"

"Quiet, quiet, stay calm man, don't panic. Tell me what happened."

"It'll be a long story that I have to tell you."

"That's okay, just start at the beginning, I'm all ears."

Not once did Jonathan interrupt him, while he recounted what had happened to him in that deep trance.

After he finished telling his story, the shaman began to talk. "The symphony is not lost. It's in your whole being . . . in your soul . . . your mind!" We merely have to get it out."

"Hey you're right. Indeed, it's constantly in my head. Just as in my youth when a popular song stuck in my head, after I heard it frequently on the radio." A smile appeared on his face. "I will try to do something, play some of it, a sketch thus, of what's in my head—hopefully I will succeed."

He began to whistle.

"Wait, hold on!" Surprised he looked at his friend who pulled out his mobile phone. "Oh you're recording it, beautiful!" He closed his eyes and after a few seconds he started whistling again. Jonathan closed his eyes too and bobbed his head to the tune coming from Dave's pursed lips. Dave was enjoying this and kept whistling, paused, shook his head and continued whistling until the end of the symphony he heard in his head. "Bam . . . bam . . . bammm," he repeated vocally.

"Holy Manitou, as you often say Jonathan, that's it!"

"Splendid, we have recorded that nicely. We will leave tomorrow and have an expert in classical music listen to it.

"Great idea, they have to transcribe it into musical notations, etc."

"Do you know where to find such a person?"

"I have to think about it, but that shouldn't be a problem."

"The music conservatory is where we have to be!" Dave exclaimed excited.

"Maybe Ann knows someone," Dave said with a pensive look. Dave took another sip of water and started coughing.

"Sorry, I almost choked. But it only now occurs to me—when I was a little boy of eight or nine, I used to play the harmonica. The symphony will sound much better on that instrument than my whistling."

"Then, we should buy one when we get home."

"Buy? They probably don't make those anymore, but mine should still be somewhere in a box in the attic of my father's house.

"You should definitely find it. It's only a question of a thorough search."

"I will certainly do that," Dave said enthusiastically.

A few moments later he stood in front of the fridge. He was terribly thirsty again.

Two days later, despite the stranger's threats, Ann attempted to contact the two men in various ways, to warn them of the impending danger. Her email messages went unanswered. She received no response to her voice messages either. Desperately she wondered how that was possible. With today's modern communications technology, you could contact someone even if they were at the North Pole. Suddenly she remembered the strange man again. From the moment she saw him, it was clear that he was one of those enemies. "My God," she thought, "these odious creatures are, indeed, powerful and super intelligent. The fact that they are able to manipulate mobile calls and the Internet proves that clearly." She lay on her bed. Since those threats, she had allowed Jimmy to sleep in her bed. She did not lose sight of him, not for a moment. "What am I to do now?" Her little boy was sound asleep. She put her arms protectively around him, kissed him on the forehead, as her tears flowed freely. She also thought about Jim's father, who when the boy was only a year old, died in a tragic automobile accident on his way home. Her thoughts returned to the dream in which the mysterious *Nin-khursag* had appeared, whence she always drew strength and optimism. Ann felt calm again and took a last look at her boy, before closing her eyes to fall asleep an hour later.

It would be a night she'd never forget . . .

TWENTY-FOUR

The van had already been under way an hour and a half, when the engine began sputtering then stopped. The driver started cursing aloud in Spanish. It was noisy inside the van when the passengers began talking past one another. Dave awoke from a nap, looked at Jonathan quizzically. "Something is wrong with the motor. It suddenly stopped." Dave yawned loudly: "What now?" Jonathan shrugged. "The driver went outside to take a look."

A total of eleven people, the driver included, were inside the van, a blue sprayed vehicle with a dented bumper. Some of the passengers had remained seated undisturbed. One was reading a newspaper and Dave noticed another one smiling, as he leafed through a Playboy magazine from years back.

The driver yelled something in Spanish to the man sitting in his seat, trying desperately to get the engine running again. The driver who was busy fiddling with the motor, yelled again at the man, who made a new attempt.

It did not work.

The driver came back inside the vehicle with a tired face and yelled "attencion!" "It looks like he doesn't have good news for us," Dave said. The man said a few things in Spanish that were met with a murmur of protests and screams.

"What's he saying?" Dave asked the shaman. Before the latter could say anything, the driver said in English with a heavy Bolivian accent: "My battery is dead. We have to . . . ahem wait and hope another vehicle will come in this direction to help us."

"And if that does not happen?" A tall skinny white guy, with a straw hat on his head, shouted in English.

"Well, quite simply, we must spend the night here," was the response.

The passengers started yelling past one another. The driver shrugged. Someone shouted in English . . . "this is too crazy for words man, who drives a bus here without a spare battery?"

"The battery is as good as new, I bought it only three months ago, the driver moaned. "Go to hell man, you're full of shit, you're lying. I don't believe it," a smiling young Bolivian yelled in English. After a while silence reigned in the van. Outside a few men were smoking and talking softly. Occasionally there was laughter.

The driver sat behind the wheel with a somber face, drinking water from a bottle.

"Shall we go outside and stretch our legs?" Dave asked his old friend, who nodded. When they were outside, they noticed that it would be evening soon.

A sad moon, through some dark clouds, illuminated the Pampa which looked desolate and lonely. A number of stars appeared in the dark sky. The shaman grabbed his pipe. As he packed it with tobacco, he said: "Another delay, we are losing another day, while the horror from that side is steadily increasing." He gestured with his pipe to the sky. Dave nodded. "Strange thing with the battery, can you guess what I'm thinking?"

"It's possible," said the shaman. "This foe is beyond us in any field. It won't surprise me if the battery has been tampered with."

"How so?"

"Manipulation of electricity."

"What do you mean?"

"What do people really know about this phenomenon?" We can generate and make good use of it, but explaining what it is exactly, is still a mystery to the greatest engineers and scientists. And as to the origin of it, we are still left in the dark, all such theories notwithstanding."

It started to get pretty cold. Dave shivered beneath his hooded sweater. "Hey what's that?" said Jonathan, pointing in front of him. Far away from the place where they were standing, two faint lights appeared. "Are those people coming over there?"

"People on foot in the Pampa? I doubt it," Jonathan said, while taking a closer look. The flickering lights drew closer. The other men had noticed the phenomenon also and made observations and comments. Just as sudden as the lights appeared, they disappeared again. Suddenly the wind began to blow. The wind increased in volume and blew fine sharp sand into the air. They felt it tickle on their faces. "Inside!" The shaman ordered. They fled back inside the bus, where the other men, who previously stood outside, were already seated. The wind increased in strength and the van began shaking in a strange way, by the sheer force of the rising storm. "How can that be?" screamed the English-speaking man, above the din of the other voices and the squeal of the forces of nature outside. The van started shaking back and forth. "In the Pampa it's always dry and often windless, this is not normal in this region." He sat in front of Jonathan and Dave. His eyes looked wild and frightened.

Then it happened. The van began to tilt and as if that was not enough, the vehicle was lifted, like by a giant hand and cast into the air. The passengers started screaming. Then the van fell down again on the rocky ground. The blow was hard and deafening.

The last thing Dave noticed, before passing out, was the door sliding open. He hit his head very hard against the glass on his right.

Ann was barely asleep, than she awoke again. She felt as if the house began to shake violently. When she opened her eyes, she noticed nothing. "Oh my imagination is playing tricks on me", she thought and closed her eyes. Suddenly the house began to vibrate slightly again. "My God, there it is again. What is that?"—she thought anxiously. From behind the bedroom door she heard softly many surprising sounds: whispering, panting which had begun spreading throughout the house. Tiny sweat drops appeared on her face. She got very cold, suddenly. "No! This is not my imagination." She listened more attentively. The whispering and panting slowly ebbed, but the vibrations returned with increased intensity. The lamp on the nightstand shook back and forth.

She had a strong feeling that something frightening was approaching her bedroom door.

She took Jimmy in her arms. The little boy mumbled something in his sleep and suddenly opened his eyes . . .

Instantly, he seemed wide awake.

His eyes were firmly fixed on the door.

"Mommy, there is bad thing there." He pointed his little finger at the door.

"Oh, it's nothing, go back to sleep, baby." Her words were belied when the vibrations of the house, which had ceased briefly, began again. The vibrations turned into violent shakes. "My God, this has to be an earthquake." Somewhere in the living room something fell to the floor with a loud thud.

She was terribly scared.

Abruptly the boy wiggled out of his mother's arms, took a few steps forward and remained standing on the bed. With wide open eyes he stared at the door again. "Jimmy!" She screamed desperately. She stretched out her arms to him.

The shaking of the house was getting progressively worse.

The house started to groan under this burden. The nightstand began shifting with a shrieking sound.

She wanted to call Jimmy back, but then something strange happened.

Disbelief appeared in her eyes.

Captivated, motionless and ashen-faced she stared at Jimmy who, with eyes wide open and arms stretched out in front of him, said something unintelligible. It sounded like a kind of "tone," she thought. His fingers moved up and down in a strange rhythm . . . again and again. She was drawn by the direction of his gaze. He kept staring at the door with a serious face. Suddenly he looked much older, she thought. "Jim . . . Jimmy . . . ," the words escaped her mouth stuttering and trembling. The vibrations slowly decreased in intensity and disappeared completely. He smiled and turned to his mother, who in utter astonishment, was at a loss for words. The boy threw himself in her arms and with a serious look he said: *"Bad thing is gone."*

A radiant smile appeared on his lips again. Surprised, she kept staring at him. *"Mommy not scared anymore."*

She sighed and did not know what to say. Then, with a trembling mouth, she said: "No sweetheart. You are a brave boy, oh my dear Jimmy." He closed his eyes again, mumbled something unintelligible and snuggled more closely against her. Slowly it occurred to her. The enemy had invaded the house to inflict terrible harm, but this little boy had successfully averted that danger.

"Unbelievable," she whispered hoarsely and perplexed. And that strange sound coming from his mouth? I only imaged that," she thought. A deep frown appeared on her forehead. She had learnt that children were safeguarded from the nightmares the adults were dealing with. A journalist had mentioned this fact in a magazine a few days ago. The article was published after a thorough investigation throughout the country. Small children, pure souls, were not susceptible to those horrors from outer space. Thus, these children were a powerful spiritual weapon against these wretches, something she understood only now.

The insight drawing across her face at that instant, clearly interpreted her thoughts. She looked at Jimmy with great admiration and kissed him on his sweaty forehead. With this child, therefore, she did not need to be afraid anymore—she hoped.

Suddenly she thought "could it be that one on the good side had helped the little boy in this struggle? Or . . ." She shuddered and tried not to think of it, then dismissed that thought.

Her thoughts moved to the encounter with the stranger two days ago. The threats that guy had made at the time—threats about harming her little Jimmy if she disregarded his advice.

Thus, spiritually children were well equipped against this evil.

Physical harm was something else! Against that kind of violence, the little ones were very vulnerable. That was a cold reality for her.

Shivering, she glanced at the clock on her bedside table. The time was 03:30 a.m.

TWENTY-FIVE

He heard excited voices and also the start of a motor, or at least attempts to do so. When he opened his eyes, Dave saw a few men leaning over him. He lay on a tarpaulin. He looked around. It was daylight. Had he been unconscious all night? Next to the van was a military vehicle. Uniformed men were working with the driver to jump start the engine with two cables connected to the military vehicle's battery. Following several failed attempts, they finally succeeded. The van's engine caught on, and after sputtering violently, it began running quietly and evenly.

He was about to inquire after Jonathan's condition, when he saw a serious-looking shaman coming towards him, accompanied by two busily talking passengers and a soldier. He sat up. His head was throbbing and hurt terribly. He touched his head and noticed a band aid on his forehead. "He just came to," he heard the Englishman say anxiously. The shaman leaned over smiling. "How do you feel?" "Fine, just that damn headache, Dave replied softly. "What the hell happened? All I remember is that it seemed as though the minibus was being lifted by something and landed very hard on the road."

"The storm was the culprit," said the Englishman. "Honestly, I really don't understand any of this. There are no winds in the Pampa, as far as I can remember." Dave stood up with some difficulty. His body was pretty bruised and his muscles ached. The van lay on its side. "We had a hard time getting you out." And those soldiers, where did they come from?"

"After we had waited for hours, a military vehicle appeared in the distance. You should have seen the men, they cheered like little children as the soldiers came closer. They helped us get the van back on its four wheels."

"Two windows, on the side where the van had landed, were shattered. They attached pieces of canvas to those areas, after cleaning up the shards."

"And what about the other passengers?"

"Three sustained minor injuries. A medic examined them and put band aids on the wounds. Look, he's coming towards us." The soldier greeted Dave in a friendly manner and asked in halting English how he was doing. After examining him again and giving him some tablets for the headache, he too stepped into the military vehicle. In a cloud of dust, the army green painted vehicle drove away.

"The soldiers did not believe it was a storm, at all. They urged the driver to be more cautious and to maintain a moderate speed."

"And our stuff?"

"Don't worry about that, fortunately there is no damage." Dave heaved a sigh of relief. The driver stepped on the gas pedal. Slowly, the rickety van began to move.

After driving for several hours without any more problems, they reached a train station and continued their journey by train. Following a flight, with a stopover in Guatemala lasting several hours, they flew to Miami.

Dave, who had tried for some time to reach Ann via his mobile phone, finally got hold of her from the airport in Miami. As he spoke to her, a mix of different expressions were visible on his face, ranging from astonishment, bewilderment, shock, disbelief and relief. Jonathan, who followed the conversation and to some degree understood what it was about, eagerly took the phone when Long handed it to him.

After ten minutes he ended the conversation with the promise to contact her again immediately after their arrival at the Northwest Arkansas regional airport in Siloam Springs.

They entered the passenger boarding bridge. "Look inconspicuously behind you," Jonathan whispered to Dave. "That guy in the grey suit has been following us since the departure hall. He keeps glancing in our direction." Dave did what he was told and said softly: "I think you're right. He matches the description Ann gave us, I think he's the same guy who visited her and made those threats against her."

"We'd better watch out," Dave said softly.

"I think these enemies are following a particular strategy, we have to be vigilant,' Dave whispered back.

"Now that their first attempt to eliminate us has failed, they are bringing out the big guns."

"I believe so too," Dave grumbled, when they sat down in their assigned seats on the aircraft. Nothing eventful happened during the flight. After being in the air for about fifteen minutes, Dave went to the lavatory. When he returned to his seat, he said to Jonathan: "That guy is seated two seats further back." He grinned at Jonathan. "He's like a robot so stiff in his seat, ready to jump into action at the push of a button."

During the taxi ride on the way to the house of Dave's deceased father, where they planned to stay, something very strange happened all of a sudden. They were driving on W. Maple Street; they had to go straight across, when the chauffeur suddenly made a turn on N. Razorback Street. The chauffeur continued driving with great speed in a direction away from their destination. "Hey man, look out! You're not going in the right direction," Dave shouted. The chauffeur pretended not to hear and accelerated even more.

"Damn it, look out!" Jonathan roared when the driver barely evaded a car in front of him.

"Look at his face," Long growled, alarmed, to Jonathan. The look on the driver's face did not please them at all. It seemed as though the man was not himself. Rigid and pale, sweat trickling down his face, the guy was staring with glassy eyes. His hands, white at the knuckles, were tightly clutching the wheel.

"My God, something is not right with this man." Jonathan thought to himself. "They are trying to enter his body and he is resisting mightily. What should we do to prevent that from happening successfully? Because if that happens, he will kill himself like a suicide killer and drive us to our death . . ."

"We must do something," Long wailed with a desperate look in his eyes. One could clearly see the battle the man was fighting against an invader. The driver began moaning loudly. His mouth was open and flakes of white foam trickled down from his trembling lips.

"Jesus, did you see that!" Dave yelled to Jonathan. Dave, who sat beside the man, made desperate attempts to take over the wheel. The driver's hands were clamped to the steering wheel like vices. "Physically, that won't work," roared the shaman. "I have an idea, but if that will work is . . ." Before he was able to finish his sentence, the car flew with enormous speed past a two-ton cargo truck, made a sharp left turn, hit a guard rail, turned on its axis and flew to the right back onto the roadway. The cargo truck slowed down with all its might.

There was a shrill and screeching sound of tires on asphalt and the smell of burning rubber in the air that was blue from smoke off exhaust pipes . . .

The taxi flew further to the right, narrowly missing two oncoming vehicles, then hitting the guard rail again with a loud bang. They were tossed about. If it were not for their fastened seatbelts, this drama could have had a completely different outcome. The airbags deployed and absorbed the heavy blow as well. The engine sputtered a bit and then shut off.

Dave gasped and was shocked when he looked at the driver's face again, which had turned ashen. The face underwent a dramatic change. His eyes bulged from their sockets, as if they were about to jump out. His mouth pulled into a demonic grin that could cause the most cold-blooded man to lose his wits. He got an even bigger shock, when he suddenly realized what was happening here. *One of them had purposely invaded the man's body, in order to slay Jonathan and him!*

He was not able to keep his eyes off the driver, who was trying in vain to unbuckle his seatbelt and pull it away from his body, while staring savagely from his bloodshot, bulging eyes.

The strength of the man was astounding.

He pulled and managed to unbuckle his seatbelt, then stretched out his hands to Dave, who sat in the seat as if paralyzed. Suddenly he felt the fingers of the man around his neck. The driver was making frantic attempts to kill him! Long was shocked by the paralyzed position in which he found himself. In a desperate effort, he tried unsuccessfully to remove the driver's hands from his neck. At that moment, Jonathan intervened and with all the strength he could muster, he struck hard with his thin walking stick the face with the demonic expression. He released two more blows to the man's head. Blood from the severely injured driver squirted into the surrounding area. Any normal human being would have been out cold at that point, but this guy kept fighting. His fingers clamped tighter around Dave's neck, who at this moment was feeling pretty suffocated. His face turned blue, his breathing was constricted in a horrible way.

"My God, this is the end . . ." it went through his head. The shaman uttered several Navajo spells and struck hard one more time, resulting in more blood squirting around.

Finally, Dave felt the man's fingers relax. With a whooping sound he began breathing again.

"Good heavens, Jonathan my good friend, you saved my life." Jonathan smiled, made a nonchalant gesture and helped him free himself from the seatbelt.

"You are covered in blood man," sounded loud the worried voice of the shaman. The ensuing silence was broken by police sirens in the distance and voices nearby. Still dazed, Dave noticed that the man's fingers continued making slow grasping movements.

The man's arms were limp at his side.

TWENTY-SIX

They stood at the hospital bed, where the taxi driver lay with a mild concussion.

The doctor allowed them to talk to the patient for a few minutes. He spoke slowly and hesitantly. "When I pulled away, nothing specific happened, but after ten minutes or so, I got cold chills all over my body. It seemed like I was being extracted from my body by a strange force. It was horrible. At times, it felt as if I didn't control the car anymore, but someone else. In the end, I knew nothing about myself anymore. I was completely unconscious. When I regained consciousness, with a horrendous headache, I wondered where I was. I looked around and realized that I was lying in a hospital bed, as you can see now. I got an even bigger shock, when I became aware of the fact that my head was swathed in bandages. Holy God, what in heaven's name has happened to me? Please tell me . . . Oh my head . . ." Dave looked at the shaman and said: "These damned miscreants are continuing their evil game to liquidate us and involve innocent people who have nothing to do with it." Jonathan hissed furiously through his teeth. "Damn it, there's no way these evil creatures will break us!" The taxi driver stared uncomprehendingly at the two men. Dave took a seat on one of the chairs and looked seriously at the man. "Let me tell you exactly what happened."

The shaman also sat down on a chair.

With disbelief in his eyes, clearly upset and horrified, the man whose name was Antonio Vargas, listened to the unbelievable things the two men at his hospital bed were telling him. Antonio heard the requesting silence, when the two men stopped talking. *Nightmares, people who were possessed. Creatures from another world, who were trying to take possession of human bodies.* His jaw dropped. "So, that's what happened to me. Holy mother of God, why me?" He moaned loudly.

"Because you were transporting us in your taxi."

"They were after us, hence," added Jonathan.

"Don't worry you hear, it will be okay. I will ask the doctor when you can go home."

After an hour they left the hospital. Jonathan's thoughts returned to the moment the police and ambulance arrived. The two men had great difficulty convincing the policeman of what had happened. With an expression of total disbelief on his face, the policeman made the official report, after they answered the questions the best they could. Following their treatment at the hospital, Dave and Jonathan had to report back to the police station, where they were introduced to two FBI agents. Bill Perry and Mike Jones informed them that they were members of a task force that, for some time, had been investigating the incidents of nightmares and possessions numerous people and prominent figures in the country had been dealing with. This team would report daily to the Bureau Chief Clyde Perry, who in turn, briefed the President of the United States. Since a few weeks the president had been briefed in detail about this enigmatic and intriguing phenomenon.

"So, they want you dead, because you hold the key to ending this terrible thing?" One of the FBI agents asked, after having subjected the two friends to a long cross-examination.

"What I just told you is entirely up to your assessment," said Jonathan with a serious expression on his face.

"It sounds incredible, I know, but you have to take us on our word," Jonathan said.

"I do believe you," was Mike's reaction, but the expression on his face belied the meaning of those words.

"One moment," he said as he got up and gestured to his colleague. They withdrew to the side to deliberate. After five minutes they returned.

"And what's the next step?" asked the FBI agent.

"We have to find an expert to listen to the symphony; that person should record it and put it in musical notes immediately; then have it broadcast nationally via radio, television and in public by orchestras across the country." Juan tugged his moustache and sighed.

"And you think it will help to . . . uh . . . eliminate this horrific phenomenon?" "We have revealed to you what exactly is going on and how far we have traveled to get behind these truths," Dave said. Fatigue was evident in his voice. "Do you have an expert in mind?"

"Frankly, not yet, but that shouldn't be a big problem."

Mike looked at him seriously. "When will we hear from you when it is so far? Delaying any longer is not desirable in this situation."

"It's now two o'clock. Might we call you in one hour?" Dave asked.

"Agreed!"

"Ann called," Dave said to Jonathan, who just arrived home with some groceries from the nearest grocery store down the street. "She has found an expert in classical music. Dr. Walter Burns, former conductor of the New York Philharmonic."

"Excellent, that name sounds familiar," Jonathan replied delighted. "And when will we meet this man?"

"Tonight. He seemed immediately enthusiastic, when Ann told him briefly what she needed him for."

She had come across his name in a magazine and could still remember that he was spending his golden years in a beautiful red brick house just outside of Siloam Springs. He was no longer active in the orchestral world, but occasionally he would conduct private sessions for interested conductors and composers.

"Great, that sounds good. Hey, in the newspaper I just purchased, I read a short article about a white man in a grey suit, who dropped dead near the place where we took the taxi. The doctor who was summoned declared that the man died of a heart attack. That got me thinking. The description matches that of the guy who was following us."

"I am beginning to understand. One of them used the body of the man and then that of the taxi driver."

"These monsters are willing to go to any length to eliminate us . . . it's awful."

Dave telephoned the FBI agent again. When he hung up, he said that they were expected at FBI headquarters in Washington, DC the next afternoon. "Well, it promises to be a busy day tomorrow," Jonathan said.

"I will book three seats with the airline."

Jonathan nodded in agreement. "This case is finally going somewhere. Let's hope we don't run into any more problems in the meantime, because you never know."

They were welcomed by a tall man with grey hair and a thin moustache. He was dressed in a white shirt and blue jeans, with brown sandals on his feet. Walter Burns looked at him seriously. Thus, you have the symphony in your head, so to speak."

"Mr. Burns, I whistled it for my friend here," Dave said shyly. "We have recorded it on my mobile phone. Later on I remembered, when I was a little boy, I bugged my father to buy me a harmonica. I received it as a birthday present and you know how that goes. I practiced day and night and after a few weeks, I was able to play extremely well all kinds of popular tunes of that time. After a few months, I stopped. I had discovered a new hobby, basketball. Every now and then, if I felt like it, I would pick up my harmonica to play a few songs. When I attended college, I left it with the other stuff from my youth in a box in the attic. The day before yesterday, after all those years, I retrieved it and played the symphony from beginning to end. I did so a few times and also recorded it. I will play it for you and, hopefully, it meets with your approval, as I have not played the harmonica in a long time, but hope it still sounds okay!"

"I'd rather you play it live for me. Do you have the instrument with you?"

"Yeah, I brought it with me. From his coat pocket Dave took the instrument that had clearly withstood the test of times. "Hey, that's a *Hohner*, a well-known brand of yesteryear," Burns said surprised.

Dave had to admit to himself that he was a little nervous, for the simple reason that Burns was a connoisseur of classical music, and what if the man would laugh at his performance. Oh well, that's not important right now. He put the instrument to his lips and closed his eyes. The prof listened with a curious look and bobbed his head along to the sounds.

The reaction of Burns was surprising, two minutes into Dave's performance. "Hey, that sounds very familiar! Don't stop, carry on . . . !"

When he finished, the maestro looked up in amazement.

"Unbelievable! Young man, do you know what you just played?" it sounded hoarse from Burns's mouth. Now it was time for Dave and the shaman to look surprised.

"I don't quite understand."

"That's Tchaikovsky's unfinished symphony, also known as Symphony number seven . . . After all this time, it is really complete, as I believe, the master would have liked . . . and also on such a simple instrument . . .

hahaha . . . like the harmonica." The maestro shook with laughter. The two men and Ann laughed along.

"This is so very unique! It's too crazy for words!"

"Tchaikovsky's unfinished symphony? Please tell us more about it," Ann asked curiously.

Burns stopped laughing, grinned and cleared his throat. "Tchaikovsky composed it in 1892. He was not happy with it and, therefore, did not complete it. Based on some sketches, Sergei Bogayrjev compiled a detailed version of the symphony. In 1962 it was released on a vinyl record in the implementation of the London Philharmonic Orchestra conducted by Eugene Ormandy. I have that record somewhere in my collection. To clarify, when a solo instrument or a group of distinct solo instruments occur in a work, it is called a concert or a concert symphony concerto grosso. Thus, what you did just now is called a solo. It seems daunting to add the other instruments, in order to complete it, as it was originally intended by Tchaikovsky, had he not put it aside." Burns looked serious.

"I'm going straight to work. I am going to start with the completed part. Let me re-record it on my own recorder."

He got up and took an old fashioned MD recorder from a cabinet drawer. Burns stood still for a moment and shook his head. "No, I won't re-record anything, just play it one more time and I will record it instantaneously."

He pushed the recording button and checked the recording mode. "Okay Dave, are you ready?" Dave nodded enthusiastically and this time he played the stars off the heavens, or so he thought.

The next day when Ann, an early riser, woke up and turned on the radio for the news, she got the shock of her life. She could not believe her ears and turned up the volume. The newscaster reported the death of Walter Burns, a well-known former conductor of the New York Philharmonic. His housekeeper found his lifeless body in the study of his home very early that morning. She had telephoned the police immediately. According to the Medical Examiner, the man had been murdered three hours earlier. How the perpetrator had entered the house, was still a mystery. There were no signs of forced entry. Seriously alarmed, Ann telephoned Dave right after she heard the news.

"Hello," said a sleepy voice at the other end of the line.

"Have you heard the news already," she said quite upset.

"Which news?" Dave replied as he blinked sleepily.

"They killed Walter Burns!"

"What?" said Long, instantly wide awake. With horror he listened to Ann, while immediately turning on the radio. "My God, this enemy stops at nothing," he moaned upset. "You know what? I will wake up Jonathan, hit the shower and come over right away. I'm going to call the police also."

The policeman looked up in surprise when they entered. "I just placed a call to you to come over right away," he said.

"What exactly happened?" A distressed Jonathan asked worriedly. According to the investigation so far, it seemed that he was killed in the early morning."

"How was he killed?" Ann asked in a timid voice.

"He was killed by strangulation," the policeman replied bleakly.

"My God, and the perpetrator, do they have any leads?"

"That's so strange. There is no sign that anyone could have invaded the house. But the investigation is ongoing, hopefully later in the day we have more clarity in this murder."

"We were visiting with him, to have him listen to the symphony. He would work on it straightaway," Ann said. "I know all about that," the policeman grumbled. The FBI agent filled me in."

"Good heavens, the symphony! Where do we go from here?" Dave exclaimed worried. He was clearly struggling to keep his emotions under control. "The symphony, he was planning to work on it that very night. Have you found anything, in writing, anything . . . ?"

"We didn't find anything of the kind," the policeman replied seriously.

"What do you mean?" He looked nonplussed at the policeman.

"Nothing was stolen from that home," the policeman said.

"At any moment now, I expect the FBI, who will take over this entire investigation."

TWENTY-SEVEN

According to the investigation, the perpetrator of Burns's murder had himself locked inside the house and waited for an opportune moment to commit his crime. The keys to the house were missing. It also appeared that the MD recorder and the CD with Dave's harmonica performance of the symphony were nowhere to be found. The three friends were at Dave's home, brainstorming about how to proceed. It was nine o'clock in the evening. Dave's face wore a troubled frown. "What do we do now? Tomorrow we have to report to FBI headquarters in Washington, DC. Those people are interested in the symphony and now, the man who was supposed to finish it is dead."

"It's a race against time," said the shaman.

"The enemy's attacks are becoming worse and more audacious. Where will we find another music expert, who would want to do this job?" Ann said, as her hands played nervously with Jimmy's hair, while he looked around with big wondering eyes.

"Finding such a person shouldn't be a problem. But he must be constantly protected against them," Jonathan said with a grim expression on his face. He exhaled. "Let me think for a moment."

"Hey guys, I have an idea!" Surprised the two men looked at her.

"Jimmy," she said softly.

"What do you mean, your son?" Dave asked surprised. A big smile appeared on Jimmy's face when he heard his name. Curious, he raised his head and looked at his mother. "I did tell you how he protected me that night."

"Yes, but how in God's name do you expect him to protect a musician who is working on a symphony?"

"Very simple," she replied resolutely. "Jimmy has to stay near that person."

"Thus, very close," Dave said rubbing his chin.

"In the same house and also in the same room," said Jonathan, with a stern look.

"That's what I mean," Ann said with a sigh, as she played with her son's hair. Jimmy sat there as if he understood what they were talking about; his eyes looked seriously at the two men. "Do you mean that he will be alone with that person? This kid is like any other child, full of energy; how could he sit quietly . . . and . . . ?" "Of course I will be there too, I don't intend to leave my son there unattended," she interrupted indignantly.

"We have no other choice," said Jonathan.

"Alright, that's what we have to do then. And now we have to hurry and find another musician to finish this work," Dave mumbled. "Luckily, the public has no knowledge of what's going on behind the death of Burns," Ann said cheerfully.

"The FBI requested that the police keep this under wraps for the time being," Jonathan remarked. "Hey, that gives me an idea. I will telephone that FBI guy and request that they urgently find someone for us." As he said so, Dave took his mobile from his coat pocket. "Hold on a second. The enemy is scientifically far beyond us. They keep an eye on us and are definitely bugging our phones," Ann said worried.

"The one, what's his name again?" A puzzled expression appeared on Dave's face.

"Mike," Jonathan said softly.

"Yes, he's still in town. He was also planning to leave for headquarters tomorrow. Give him a call and do not tell him what it's about. We will tell him when he gets here."

"I will think of something, Dave replied, while dialing the number.

The FBI agent did not need many words to understand what it was about. He frowned. "I will get on it right away. I will call you back in half an hour. The code is—*a canary caught*. Then you will know that we found someone. Please come immediately to the hotel where I'm staying." He took out a pocket diary, tore out a sheet, quickly jotted down something and handed it to Ann.

"We have to be extremely careful," Mike said.

He left without saying goodbye.

Completely baffled, Ann read what was written on the piece of paper: 12 W. Dickson Street.

"But, there is absolutely no hotel there!"

"How do you know that exactly?" Dave asked.

"I know, because I take that same street on my way home."

"Oh yeah," he responded surprised.

Jonathan gently hit his fist on the table. "Everything is messed up now. I immediately noticed something odd, when I looked at that guy. At the interrogation the other day, he spoke completely normal, but just now, he stuttered ever so slightly, you know what I mean."

"One of them succeeded in taking possession of Mike's body." Jonathan cleared his throat.

"What struck me was his inflexibility. Apparently, they have no feelings resembling human sentimentalities," he said softly. "Feelings such as greetings," remarked Ann. "Damn! It's driving me crazy. They have infiltrated everywhere. They know everything about us. Who knows how many people in our society have become victims of these . . . nefarious creatures." Dave looked distraught. Jonathan scratched his forehead. They surround us . . . they can waylay us; they are watching us. We must always be on the qui vive—even more than that."

"I think we're all alone now. We cannot trust anybody—Jesus I'm getting sick of it." Dave stood up, staring blankly for a while and sat down again.

"It is dreadful for those who were driven out of their body. Those creatures are using the brain of their victims to think and their mouth to speak." He took a deep breath and held on tightly to the back of his chair, his white knuckles betrayed the tension he felt.

"What should we do next?" Ann intervened. Dave shrugged despondently. The only one, who sat there quietly, was the little boy, playing with his teddy bear, oblivious of what was going on in the room.

All of a sudden there were soft sounds on the roof and the windows.

"Rain," said Jonathan.

He looked at the clock on the wall. "Midnight."

The rain howled in the air and bounced against the windows. "The Weather Channel had forecast this yesterday," Ann declared.

"But this kind of rain and wind storm are unusual for this time of year," Jonathan grumbled surprised.

"Look man, everything on this planet is being disrupted, since those damn wretches have showed up," Dave snorted.

He yawned and stretched. "Don't you have to put the little one to bed?" Ann leaned back in a leather armchair as she watched her little boy. "You're right, come Jimmy, it is way past your bed time." Jimmy climbed in his mother's lap, snuggled closely and closed his eyes.

TWENTY-EIGHT

The images on the television screen during the late news were shocking. The number of suicides had reached an incredible climax, the news reader reported with a straight face. The most tragic of all this was that many people were afraid to fall asleep, leading to a complete breakdown of society. The economy became hopelessly stagnant. Train and airplane traffic came to a complete halt, since most people were afraid to leave their home.

In the late news they also showed pictures of escalating church services, which had backfired on churchgoers seeking reassurance for these serious unprecedented problems in their lives. Problems, which every day and night were getting worse.

Ann had a sad expression on her face. She thought of all the nice and friendly people in Siloam Springs and beyond, who were threatened with eviction from their own body by those creeps from an unknown world. The knowledge that all these people with their idiosyncrasies and habits, engaged in whatever they were entitled to, from loving one another, quarreling with one another to doing anything they felt like doing, were now threatened by a danger they could not understand with their conditioned mind. With the arrival of that bright star, clearly visible in the nightly sky, the rediscovered planet *Nibiru* was leaving increasingly deep scars on the psyche of the *earthlings*. Knowing that she and her two new friends could reverse this terrible tragedy, filled her mind with fear—weren't they losing this fight? She also thought of the appearance of *Nin-khursag* in her dream a while back. Ann could not believe that *she* had abandoned them. Ann never dreamt of her again, although she was glad not to have any more nightmares. She had an irrational fear for those, like a bird has for a cobra. That image of those creeps, she knew for sure, would always remain with her.

"Jonathan, you told me once about an old woman both of you saw in Bolivia. What you told me then, made me think. You never spoke of that encounter again. Suddenly I remember it. I have no doubt that she was *Nin's* messenger, or perhaps *Nin* herself, disguised in this material world."

With a disappointed expression on her face, Ann said: "But it looks a lot as though *she* left us to our fate, what do you think?" Jonathan looked at her cautiously. "I don't think so. I have the feeling that *she* is working on a plan to strike her opponents a heavy blow, at an unexpected moment."

"Hopefully your sentiments come true," Dave grumbled unsure, as he restlessly and annoyed ran his fingers through his hair. "Frankly, I'm really beginning to have my doubts now."

"Theoretically, these creatures can do almost anything we humans have never heard of," Ann said disconsolately.

"You're right," said Jonathan. He began thinking aloud—she gave us the completed symphony through Dave. It's important that it be heard at a point in time still unknown to us. She said nothing about a concert or anything like that . . ."

"Hey, wait a minute!" He looked at Dave.

"You did an excellent job on the harmonica. Keep that instrument always ready, for when the time comes."

"I . . . I have to disappoint you somehow."

"How so," Jonathan looked at him anxiously.

"I'm slowly losing the symphony in my head!"

"What are you saying . . . damn! No, not that too . . . !"

They stared at him, stunned.

Silence.

Only the roar of the storm outside penetrated into the room.

Suddenly the silence was broken by the shaman's voice.

"Hey, I still have it on my mobile!" Jonathan searched his pockets. He frowned. "Holy Manitou, where did I leave it?"

"Maybe at home or in another coat pocket," Ann replied with a weary voice. "Strange, I always check whether I have it with me, but I don't recall doing so this morning, when I went out the door." Jonathan was tired. His face looked haggard; his hands were trembling as he packed his pipe with fresh tobacco.

"I foresee that what's happening now will substantially change history and also today's events in favor of . . ." Jonathan shook his head and took a deep puff on his pipe. After these words of the shaman, an oppressive silence fell upon the room. Then, suddenly hard knocks on the front door of the house and a voice outside yelling something. "Good Lord, who's that at this late hour?" Ann cried terrified. They pricked up their ears. There were heavy footsteps around the house. The knock on the door sounded louder above the roar of the rain and wind.

A voice loud and firm shouted . . . "FBI open up!"

"FBI?" Dave whispered, terrified. "The FBI agent was here only a few hours ago. I don't get it."

"Is it really the FBI?" whispered Ann as her face turned ashen, pressing Jimmy even closer against her chest.

Dave quickly walked to the door.

"Take it easy . . . hold your horses," he yelled nervously.

When he opened the door, six men in dark suits, with dangerous looking weapons at the ready, stormed inside and nearly ran him down.

"Is everything okay with you all? We're here to protect you! Where are the others?"

Severely confused, perplexed and yet relieved, Jonathan pointed backwards. The men rushed to the rear. Terrified Ann and Jonathan watched as the men entered the room.

Ann held her boy tightly in her arms. She stood there trembling, with fear in her eyes. She could hear Dave shouting from afar . . . "It's okay." She wanted to yell—what's okay, but only a deep sigh escaped her mouth. They were now surrounded by the men.

To her surprise, despite the ruckus, Jimmy remained asleep quietly with a slight smile on his lips.

A short stocky man, apparently the leader of the group, gestured reassuringly with his arm.

Dave entered the room, accompanied by a man in civilian attire. "I've come here with a special assignment from the president of the United States of America." He paused, and said something unintelligible in the ear of the leader of the group, who nodded and clasped his hands. The man in civilian clothes spoke again. "My name is Will Cameron, Head of the FBI field office in this state. I apologize for the uninvited visit, but we had no other choice."

He cleared his throat.

"In the past few hours, horrible events have taken place in Washington, DC and the rest of the country; the president narrowly escaped an attack in the White House. We are here to take you to a secret location, with the utmost urgency, but more on that later. We have to leave now."

"Now?" Ann asked in a hoarse voice.

"Yes, right now!"

"Earlier we had a visit from an FBI agent who . . ."

Before Jonathan could finish his sentence, Cameron said brusquely: "Oh Mike Jones, he was found dead not far from here."

In the pouring rain with gusty winds, they walked to a black SUV. No sooner were they seated, than the car drove off with screeching tires. They did not say anything, but waited anxiously for further developments and the things to come.

Jonathan had to admit to himself that he had a terrible foreboding—something dramatic is going to take place, I'm convinced.

The chauffeur drove the car with enormous speed; they had to hold on firmly in order not to be pitched about. After fifteen minutes the car stopped with squealing brakes.

"We have to get out now," it sounded, not particularly unfriendly. Cameron, who had spoken those words, quickly opened a door. It had stopped raining. A stiff breeze blew into the vehicle. They found themselves on a neatly mown lawn. The silence was oppressive. Occasionally, you could hear mumbling of men. A few yards further stood a helicopter. "We are leaving with this aircraft," said the leader of the group.

Ann, who was holding her boy tightly in her arms, cried out: "Where are you taking us in heaven's name . . . ," but her words were lost in the racket of the chopper's starting engine. The screeching sound of the turbines was slowly drowned out by the din of the double rotor. They quickly climbed into the chopper which, shortly after they had boarded, rose straight up.

The flight had barely lasted half an hour, when the helicopter landed on a grass field far beyond Siloam Springs. The large loading doors opened; numbed by the cold with stiff muscles, they stepped out of the aircraft. Along the way they had not talked much.

During the flight, Jimmy had opened his eyes on two occasions and had given his mother a kiss. The second time he had mumbled something like . . . "Mommy don't you be afraid . . ."

She had smiled at him encouragingly. When they drove away, shortly after boarding a van, the little boy almost jumped from her arms. This time he was wide awake and stared around. His little face looked serious—like a big boy, Ann thought. After a ten-minute ride, the vehicle drove onto a site enclosed by a fence. Will Cameron muttered something in his mobile and the gate opened. The light cream stucco building was constructed around a square courtyard. They took an elevator up. The elevator stopped on the third floor. They got off and came to a veranda that ran around the building. When Dave looked over the railing, he saw a swimming pool and a garden below. They walked a few more yards and arrived at an oak door. When Long momentarily turned around, he noticed that a number of identical doors opened onto the veranda. Cameron pressed twice on a panel. Slowly the door swung open. He signaled to follow him.

TWENTY-NINE

Via an impressive lobby lined with old paintings, they entered a large bare room. Dave's thoughts were occupied with the frustrating puzzle of what would happen next . . . where have we arrived now . . . if? He was interrupted in his thoughts, when Cameron said softly, while stroking his chin: "He's expecting you . . ." With these words, he quietly closed the door through which they had entered and disappeared. They waited a few minutes, then a side door to the room opened and an impressive, tall, burly appearance entered the room and greeted them with an extended hand. "Welcome," it sounded from the thin lips of a tall black man, an African-American type, with grey hair. He was impeccably dressed in a blue two-piece suit that fit him like a glove. His face had a blank expression; he stared at them with a meaningless gaze. Jonathan felt a shiver run down his spine when he studied the man's eyes. An unprecedented evil vibration radiated from the stranger towards him. Ann, who also noticeably sensed it, held Jimmy closely in her arms and hugged him even tighter. He struggled to get out of her arms, turned his head and looked intently at the man . . . Dave, who saw that, poked Jonathan in the side.

"Did you see that . . ." he whispered to the shaman, who only nodded.

The stranger looked at the little boy for a moment and briefly a slight grin appeared on his lips. He cleared his throat and said: "I will get straight to the point. You think you can thwart the plan, but you will never succeed . . . I have completely destroyed the final part of the symphony! And that is exactly what I will do with you too. We will completely take over this planet and you, *earthlings*, will get a place in all of this that suits you."

He paused.

"I can make life, after we have become lord and master of this planet, considerably easier for all of you present in this room. But . . . I have a price . . . ," he smiled.

He looked at Jimmy and then at his mother. "Madam . . . give the boy to me *now*! The voice of the man in blue was as the sound of death.

Ann thought—mentally he cannot harm Jim, but physically . . .

"No," she whispered, her mouth was dry with sheer abhorrence. "No . . . No . . ." She wanted to say something else, but her voice choked in her throat. Then her eyes narrowed as she screamed fiercely emotional: "Never . . . Never . . . Go back to the hell you came from, you damned creature! . . . Who do you think you are . . . monster!"

He grinned maliciously.

"Why are you so upset?"

Ann began sobbing loudly, stopped and stared at him with pure hatred.

The shaman took a step forward and started chanting in Navajo. Mockingly the man looked at Jonathan. "What do you think to accomplish with that chanting . . . ?"

Dave suddenly got a hunch and searched his pockets. He took out his harmonica and watched the shaman, who had stopped chanting and was humming now, with his eyes closed. He held his right hand stretched out in front of him, the other hand was clamped tightly around his cane.

There was more.

The little boy wiggled, with great strength, from his mother's arms. She was forced to let him go. It was as if a voice in her head said . . . *let him go . . . put him on the floor . . .* she immediately put him down.

155

The little boy stood on his bowed legs, his eyes constantly focused on the stranger, who said nothing. The man in blue cocked his head to one side. It seemed as though he was listening to something. Dave brought the harmonica to his mouth, hesitated . . . and suddenly froze in terror.

From afar a sound was audible that gradually increased in volume. Someone was playing a harmonica. *"My God . . . the symphony . . ."* Ann whispered with tremendous fear in her eyes . . .

The sound became louder.

Dave began shivering all over his body.

Intense fear appeared in his eyes and he shook his head violently several times.

The little boy averted his eyes from the man and looked at Dave. For an instant they had eye contact. Dave felt a warm sensation throughout his body. Ann continued watching, with ever increasing bewilderment, what was unfolding before her eyes.

Suddenly it seemed as if Dave shook off the fear and numbness.

With a swift movement he brought the instrument to his mouth.

The dark figure in blue took a step forward and raised his hands imploringly.

There was a foul repulsive odor around him that became increasingly strong.

The beginning of Symphony number nine emerged as a fresh downpour from Dave's harmonica. Ann clearly heard how he tried to elevate his performance, in order to ward off or drown out something. That something was the sound of the harmonica of his unknown antagonist.

Jonathan knew with certainty—if Dave could bring the symphony to a successful conclusion . . . complete it . . . that would be the end of this enemy.

The performance became feverish, hysterical . . . horrifying, but until the end, it retained a quality equal to that of a super talent. On Jonathan's face appeared triumph . . . he wanted to scream, but restrained himself.

Louder and wilder became the sound of the howling harmonica.

Ann was watching everything dumbfounded. She also knew that an indescribable fear compelled him to play like this . . .

This was harmonica virtuoso. Dave Long was dripping with sweat.

He swung up and down like a drunk.

The big lamp on the ceiling suddenly began flickering . . . on . . . off . . . on . . .

Was all this taking place here, her imagination?

It also got alternately cold and hot in the room.

The floor beneath her high heels began to vibrate. Those vibrations suddenly became stronger. It was as if the ceiling, floor and walls were under intense pressure and would rip open, with a loud bang, at any moment.

Ann felt her muscles stiffen.

She was overcome with fear . . . fear . . . fear . . . The nauseating stench of sweat and other indefinable heavy scents became stronger and began to hang like a curtain in the room.

But the worst was yet to come.

Nobody was paying attention to the toddler, who had removed himself from his mother and was walking towards the stranger, who stared at the boy full of hate.

The intense rousing music and the flickering light of the lamp on the ceiling revealed a frightening scene before their eyes. The little boy and the big man, like boxers in a ring, began cautiously circling each other.

Ann stood frozen, unable to move . . . and watched with horror in her eyes, her little son and that sinister figure circling each other over and over again, appraising each other's strength.

Suddenly, it began to dawn on her.

What was happening here rose beyond normal human concepts. In this room, an immense showdown was developing, between two daring adversaries. *It was a spiritual battle between two great forces.*

Aided by Dave's harmonica performance, bordering on madness and the ancient spiritual ancestral knowledge, engulfing and emanating from the shaman, stood little Jimmy, moving his hands.

Gesticulating strange hand movements, evoking in Ann's mind memories of that awful night in her home, when the boy expelled evil from the house. She also thought of that strange look of mysterious adult wisdom on Jimmy's face that night, which she could also see now, as she looked at the boy.

Then suddenly!

The stranger stumbled. There was abruptly a frightful bone-chilling scream . . . he fell to his knees . . . uttering terrible cries and words in an unintelligible language. Then he fell on his side, shaking violently, again and again and again.

Once more, that unbearable scream. The voice trailed off. Ann felt as if she would suffocate at any moment. Her nerves were about to give, when she noticed the shaman undergoing a complete change, as he dropped his cane.

The stranger stood unsteady, staring straight at the shaman, who let out a violent protesting groan, as he fell hard to the floor. A quick triumphant smile appeared on the mouth of the man in blue, vanishing just as quickly as it appeared.

It changed into a bitter grimace.

Quickly she glanced at Jimmy.

A mysterious pensive look shone in the eyes of the little guy.

For a moment, Ann had a strong feeling that it was not her little son standing there, but someone else . . . a higher spirit that had taken over the body of the little boy. She did not know anymore and shook her head despairingly, chewed at her lower lip and with her hand over her mouth she continued watching in horror. The ghastly, disgusting stench became less and faded slowly. Also the vibrations of the room abated and stopped.

What happened next, bordered on the incredible . . .

The shaman's body on the floor was shaking, as he moved wildly with his arms. Movements that became increasingly weak.

Then, suddenly again that blood curdling scream. Jonathan's body rapidly changed into sand-colored dust, blowing through the room like a rustling spray rain, evaporating as if hit by a gust, until nothing remained, but a small puddle of a watery substance. Ann trembled on her legs—one shiver after another ran down her spine. Filled with disgust and hatred, she shifted her gaze once more to the stranger, overcome with desperation for the dramatic loss of the shaman. The man in blue suddenly turned his head, glanced at her with a tired look and smiled satisfied. Then with a deep liberating sigh, she heard him say: *"It's over . . . this tragedy is over!"*

Ann's heart skipped a few beats, she kept staring at him terrified and severely shaken, with big teary eyes.

—No!

This cannot be!

This is not possible . . .

Her ears did not deceive her, did they?

That was the shaman's voice!

In a flash of insight, she understood instantly what had happened in those few minutes.

Trembling over her entire body, she whispered hoarsely: "My God, what a great sacrifice you made to liquidate this evil . . . !"

At that moment, Dave's harmonica performance ended—the symphony had been completed and played out . . .

EPILOGUE

Now I am on the eve of a great event and looking back to the turbulent past of this planet, it penetrates slowly unto me, the impact we have had on the thinking and actions of these simple-minded beings that had completely lost their way. As for the major environmental issues existing today, which they can no longer cope with, we have the answer. Slowly but surely, those too will belong to the past.

While we infiltrated in leading positions of major countries and positively changed the world events with tact and wisdom, we realized that somewhere far in the universe or perhaps also close by, a great intelligence determines the direction of a mighty plan spanning millennia, of which we are merely the handles to accomplishing this.

I often go back in thought to the beginning, when I shared a physical body with the spirit of a toddler named Jimmy and finally became one with the old soul reincarnated in that little child . . .

That tall, often serious-looking African-American man, frequently seen in close proximity to me, still is my principal advisor after all those years.

The recollections of my time on Nibiru are gradually fading. There will come a day when all this will forever be erased from my memory.

We acquired all of their habits, complemented with our immense wisdom which we use to make something beautiful of this magnificent blue planet.

This living organism called Earth, once distraught with what she brought forth, may now rest assured further maturation to which it is intended in this mysterious cosmic event, of which we catch a faint glimpse at times.

Our children and grandchildren will ever behold truths, dismissing all the dogmas of the elders as jokes invented by those who do not know any better.

The planet of my origin has been slowly disappearing, once again, in the shadows of Neptune, after having completed the cycle which brought it close to Earth. Only a reminder will remain of what took place, when mankind made a leap from the dark ages of war, hunger and disease, but that too shall fade one day and be reduced to a misunderstood dark legend.

Tomorrow I will be inaugurated as president of this great nation, which I will propel to greater heights . . . and I know that she who loved very much, what she created so long ago, will now be pleased that her spirit in the coming centuries will continue to watch over her legacy, until mankind—to the extent it has spiritually evolved—will shape the plan ever more without her help.

Whatever the ultimate shape will be is only known to that higher mysterious universal intelligence, many call GOD.

THE END

ACKNOWLEDGEMENTS

My inspiration for writing this story came from Sir Laurence Gardner, Howard Philips Lovecraft, Edgar Allen Poe, August William Derleth, Charles Charroux, Eric Von Daniken and Robert N. Monroe.

Special thanks to Vera Hupsel.

Thanks also to Jaïr, Chanel, Linda and Merita for your support and inspiration.

I dedicate this book to my children Priscilla, Rodrigo and Genti.

Cynthia Joeroeja, who passed away, your spirit was with me while writing this book.

www.ingramcontent.com/pod-product-compliance
Lightning Source LLC
Chambersburg PA
CBHW021542200526
45163CB00014B/810